Direct Operating Costs of

Markus Rehsöft

Direct Operating Costs of Aircraft Fuel Systems

System Architecture, Analyzing Methods, Contribution of Components

VDM Verlag Dr. Müller

Imprint

Bibliographic information by the German National Library: The German National Library lists this publication at the German National Bibliography; detailed bibliographic information is available on the Internet at http://dnb.d-nb.de.

Cover image: www.purestockx.com

Publisher:
VDM Verlag Dr. Müller Aktiengesellschaft & Co. KG, Dudweiler Landstr. 125 a, 66123 Saarbrücken, Germany,
Phone +49 681 9100-698, Fax +49 681 9100-988,
Email: info@vdm-verlag.de

Produced in USA and UK by:
Lightning Source Inc., La Vergne, Tennessee, USA
Lightning Source UK Ltd., Milton Keynes, UK
BookSurge LLC, 5341 Dorchester Road, Suite 16, North Charleston, SC 29418, USA

ISBN: 978-3-639-03122-5

Table of Content

List of Pictures

List of Tables

List of Symbols

a	Opposite dimension
α	Angle
AMC	Airframe Maintenance Costs
c	Hypotenuse
C_I	Cost of a Delay of up to 29 min.
C_{II}	Cost of a Delay between 30 min. and 59 min.
C_{III}	Cost of a Delay of equal or more than 60 min.
C_C	Cost of a Cancellation
CAC	Cabin Crew Costs
CPC	Cockpit Crew Costs
D_I	Probability for a Delay of up to 29 min.
D_{II}	Probability for a Delay between 30 min. and 59 min.
D_{III}	Probability for a Delay of equal or more than 60 min.
D_C	Probability for a Cancellation
$Delay_{SYS}$	Delay and Cancellation Costs of aircraft systems
DEP	Depreciation
$Depr_{SYS}$	Depreciation of aircraft systems
DMC_{SYS}	Direct Maintenance Costs of aircraft systems
DOC_{SYS}	Direct Operating Costs of aircraft systems
$DOC_{SYS,ext}$	Extended Direct Operating Costs of aircraft systems
EMC	Engine Maintenance Costs
FUE	Fuel Costs
$Fuel_{mf}$	Fuel Costs due to fixed system mass
$Fuel_P$	Fuel Costs due to power off-takes from the engines
$Fuel_{SYS}$	Fuel Costs of aircraft systems
$Fuel_X$	Fuel Costs due to physical cause X
$FuelPrice$	Fuel Price
g	Earth Acceleration
γ	Flight Path Angle
i	Flight Phase
INS	Insurance
INT	Interest

k_p	Ratio between a thrust specific fuel consumption with and without power off-take and the related power off-take divide by the engine's take-off thrust
LAF	Landing Fees
L/D	Lift to Drag Ratio
LR	Labour Rate
$m_{A/C}$	Average Aircraft Mass
$m_{i,x}$	Mass in Flight Phase i
$m_{(i-1),x}$	Mass in Flight Phase $(i-1)$
$m_{fuel,x}$	Mass of fuel consumed due to a cause X during a flight
$m_{fuel,i,x}$	Mass of fuel consumed due to a cause X during a flight phase I
MC	Material Costs
MMH_{off}	Maintenance Man Hours off Aircraft
MMH_{on}	Maintenance Man Hours on Aircraft
$MTOW$	Maximum Take-Off Weight
$MZFW$	Maximum Zero Fuel Weight
n	Number of Engines
N	Depreciation Period
NAV	Navigation Charges
NFY	Number of Flights per Year
P_i	Power taken off in Flight Phase i
$Price$	Price of aircraft systems
$Residual$	Residual of aircraft systems
SHC_{SYS}	Spare Part Holding Costs of aircraft systems
SFC	Thrust Specific Fuel Consumption
t_i	Duration of Flight Phase I
$T_{T/O}$	Take-off Thrust of one engine

List of Abbreviations

APU	Auxiliary Power Unit
CEA	Cost Efficient Aircraft
CG	Centre of Gravity
DMC	Direct Maintenance Costs
DOC	Direct Operating Costs
ETOPS	Extended Range Twin Engine Operation
FCMC	Fuel Control and Monitoring Computer
FCMS	Fuel Control and Monitoring System
FH	Flight Hour
FQI	Fuel Quantity Indicating
HTP	Horizontal Tail Plane
IOC	Indirect Operating Costs
LCC	Life Cycle Costs
MLI	Magnetic Level Indicator
MLIH	Magnetic Level Indicator Housing
MMI	Manual Magnetic Indicators
MTOW	Maximum Take-Off Weight
MWE	Manufacturer's Weight Empty
MZW	Maximum Zero Fuel Weight
OI	Operational Interruptions
SMA	Single Motor Actuator
TLAR	Top Level Aircraft Requirements
TMA	Twin Motor Actuator
UERF	Uncontained Engine Rotor Failure

1 Introduction

"Airline operators are increasingly demanding of new products in a highly competitive market, where even small margins are hard won." (**Rogers 2002**). Thus an aircraft manufacturer, needs to exploit its in service support experience and feed it back into the design of future products to achieve the ever more demanding requirements of its customers.

In the early stage of aircraft design, the manufacturer normally uses comparisons between several configurations by using Top Level Aircraft Requirements (TLAR) like Maximum Take-Off Weight (MTOW), Manufacturer's Weight Empty (MWE) and Direct Operating Costs (DOC) per seat and range. Also when considering the configuration's effect on Life Cycle Costs (LCC), it is getting more and more important to recognize "…both the manufacturer's and the operator's profit." (**Bradshaw 2004**).

After an academic literature search a method called DOC_{SYS} has been found, which uses system specific parameters and TLAR to estimate system DOC. **Bradshaw 2004** has successfully demonstrated the coupling of DOC_{SYS} with DecisionPro™ for a notional system. The latter is a tool used historically by the Financial Services sector for scenario modelling and strategic analysis.

DOC_{SYS} has been used to consider system and aircraft specific parameters and DecisionPro™ used to evaluate uncertainties within this approach. This combination should give a first impression of the main cost drivers of a Fuel System and the sensitivity of DOC to imperfect data and which data have the strongest influence on this formulation for a given degree of uncertainty.

2   Life Cycle Costs and Operating Costs of an Aircraft

As described in the introduction, this thesis deals with an investigation of the DOC associated with a Fuel System. Before an investigation starts, it should be clear how DOC could be defined and what impact it has on the whole cost structure through the lifetime of an aircraft, the LCC.

Figure 1 shows a possible cost distribution of an aircraft's life cycle as presented by **Roskam 1990**.

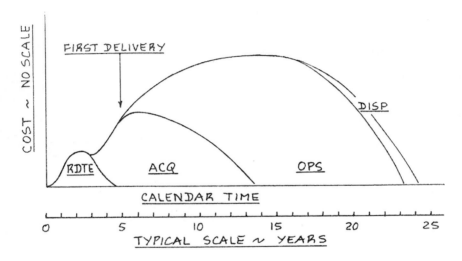

Figure 1:   Schematic representation of aircraft's Life Cycle Costs

The abbreviations in this figure describe four main phases during an aircraft program, where:
- RDTE is the research, development, test and evaluation phase
- ACQ is the acquisition phase, in which the aircraft is manufactured and delivered
- OPS is the operating phase
- DISP is the disposal phase

Related to all these phases is a certain amount of cost. Thus **Roskam 1990** models the LCC using a simplistic equation:

$$LCC = C_{RDTE} + C_{ACQ} + C_{OPS} + C_{DISP}$$ (1)

But Figure 1 does not only show the different contributions. It also gives an idea of the relative costs in each phase, indicating that the operating cost C_{OPS} makes the greatest contribution. Therefore it is important to consider operating cost especially in the conceptual and preliminary phases, which "...are responsible for locking in most of the Life Cycle Cost of an airplane!!!" (**Roskam 1990**).

Roskam 1990 splits C_{OPS} into DOC and Indirect Operating Cost (IOC). This thesis will only investigate the influence on the DOC and so the IOC is not further considered.

DOC for a whole aircraft can be modelled as the sum of price dependent, flying and maintenance costs. **Airbus/University of Bristol 2002** defines these costs as follows:

Price dependent costs:
- Depreciation (DEP)
- Interest (INT)
- Insurance (INS)

Flying costs:
- Fuel Cost (FUE)
- Landing Fees (LAF)
- Cockpit Crew Costs (CPC)
- Cabin Crew Costs (CAC)
- Navigation Charges (NAV)

Maintenance costs:
- Airframe Maintenance Costs (AMC)
- Engine Maintenance Costs (EMC)

On this basis, the DOC of an aircraft could be calculated as follows:

$$DOC = DEP + INT + INS + FUE + LAF + CPC + CAC + NAV + AMC + EMC \qquad \textbf{(2)}$$

3 DOC_{SYS} – Direct Operating Costs for Aircraft Systems

Chapter 2 shows the various contributions to aircraft DOC and the parameters that influence these contributions. The question is how does an aircraft Fuel System affect the DOC.

Scholz 1998 presents a method called DOC_{SYS} for estimating DOC caused by an aircraft system. DOC_{SYS} breaks down the system related DOC in three main parts, considering several system and aircraft specific parameters. Similar to the DOC method of **Airbus/University of Bristol 2002**, DOC_{SYS} also links the DOC to price dependent and technology related costs.

The main cost elements of DOC_{SYS} are:
- Cost due to depreciation $Depr_{SYS}$
- Cost due to fuel burn $Fuel_{SYS}$
- Cost due to maintenance DMC_{SYS}

An accumulation of these costs will conduct to an equation for the system or system component related DOC per year, which should be named like the method itself, thus:

$$DOC_{SYS} = Depr_{SYS} + Fuel_{SYS} + DMC_{SYS}$$ (3)

Beside this main equation there also exists an extended form by considering:

- Cost due to delays and cancellations $Delay_{SYS}$
- Cost due to spare part holding SHC_{SYS}

Scholz 1998 defines the extended DOC_{SYS} equation as follows:

$$DOC_{SYS.ext} = Depr_{SYS} + Fuel_{SYS} + DMC_{SYS} + Delay_{SYS} + SHC_{SYS}$$ (4)

The following chapters briefly explain how the five contributions are defined and calculated.

3.1 Depreciation

Like the depreciation for a whole aircraft, this value is time dependent and consists of a price of the system, a certain residual and a depreciation period N in years.

$$Depr_{SYS} = \frac{Price - Residual}{N}$$ (5)

The residual could be assumed as a fraction of the price, which presents the value of the system after N years. In this study the residual should be 10% of the price and the depreciation period N should be 15 years.

However, the definition of price in general is a little bit more complex. **Roskam 1990** defines a price as follows:

$$Price = Cost + Profit$$ (6)

"How cost, price and profit are viewed depends on which position is occupied in the economic process." (**Roskam 1990**). In the case of an aircraft system, the aircraft manufacturer buys a lot of system components from vendors and installs them in the aircraft, which will be sold to the operator. Thus there exist three major positions within the economic process as follows:

4

- Position of the vendor:

  The vendor has to develop and produce the required system component, which means a certain cost to him. He will then sell the component for a certain price, which is normally higher then the cost, to achieve a certain profit.

- Position of the aircraft manufacturer:

  The aircraft manufacturer has to develop the system, to buy or in some cases to produce the system components and to install the system in the aircraft. All procedures are causing costs for the aircraft manufacturer. The aircraft and the system within the aircraft will then be sold for a certain price to the operator. The price paid by the manufacturer is likely to be different to the price paid for the components individually by the operator.

- Position of the operator:

  The operator has to pay the price of the manufacturer's aircraft to purchase it. He has then to consider this price within the depreciation and thus the operating costs to fix his ticket prices to make a certain profit.

With this knowledge an investigation of the DOC would normally use the system price, which the operator has to pay for. But the above statements also show that the system price depends on several manufacturers' costs and a manufacturers' profit.

The profit is very difficult to evaluate, because the manufacturer sells a whole aircraft to the operator and therefore makes profit at the whole aircraft level. Which fraction a particular system takes in this profit would need a longer investigation for itself. Due to the time limit of this thesis, no account was taken for the latter costs.

Hence the depreciation used in DOC_{SYS} will only be calculated with purchasing prices of the system components, which the aircraft manufacturer has to pay for. This approach contains several uncertainties due to additional cost to the manufacturer, complex price commitments by the vendors and commercial influences.

Of course there are the technical requirements, which are fixing the price. In other words, the component has to achieve specifications, which maybe make the component more complex and expensive.

But there are also economical circumstances, which influence a component price. The vendor has to consider several market relevant parameters to plan his production line and to fix his profit. These parameters could be:

- The number of components installed in one aircraft
- The number of aircraft that will be produced (fleet size)
- The time (period) in which the fleet will be produced
- The number of competitors
- The relationship between vendor and aircraft manufacturer
- Price reduction due to a discount, e.g. the vendor will sell a component for than cost, if the aircraft manufacturer also buy other components at a certain price

The price data used in this thesis represents prices of one particular market situation and therefore the above statements will not be further considered.

Due to commercial sensitivity, all price and cost information in this thesis is presented in non-dimensional considerations.

5

3.2   Fuel Costs

The costs due to fuel burn are considered within DOC_{SYS} by their physical cause. "This approach helps to pinpoint the origin of fuel costs and allows to effectively find measures to reduce fuel consumption."(**Scholz 1998**).

The physical causes of fuel burn and the related costs are defined as follows:

- $Fuel_{mf}$ fuel costs due to transportation of fixed mass
- $Fuel_{mv}$ fuel costs due to transportation of variable mass
- $Fuel_P$ fuel costs due to mechanical power off-takes from the engines
- $Fuel_B$ fuel costs due to bleed air off-takes
- $Fuel_R$ fuel costs due to ram air off-takes
- $Fuel_D$ fuel costs due to additional drag caused by the presents of aircraft systems, subsystems, or single parts

Due to the nature of the investigated Fuel System, only fuel burn due to fixed mass and power off-take will be considered in this thesis. The reason, why only these fuel fractions will be included, should be clearer after the explanation of the Fuel System architecture later in this thesis.

With this knowledge, the fuel cost due to a Fuel System can be calculated with the following equation:

$$Fuel_{SYS} = Fuel_{mf} + Fuel_P \qquad (7)$$

Scholz 1998 defines the fuel cost for every physical cause X as:

$$Fuel_X = m_{fuel,X} \cdot FuelPrice \cdot NFY \qquad (8)$$

Where:
$m_{fuel,X}$ is the mass of fuel consumed due to a cause X during a flight mission
$FuelPrice$ is the fuel price
NFY is the number of flights per year

The fuel price depends on the price of crude oil, which depends heavily on world political and economical circumstances and the fuel price used in this thesis was taken from **Hume 2004**. There are 700 flights per year assumed for this study.

The question is how the several fuel masses can be calculated. To do this DOC_{SYS} divides a flight into seven phases i, where:

$i=1$ is engine start
$i=2$ is taxiing
$i=3$ is take-off
$i=4$ is climb
$i=5$ is cruise
$i=6$ is descent
$i=7$ is landing, taxiing & engine shut down

Generally the fuel consumption during a flight phase is given by **Scholz 1998** with the following equation:

$$m_{fuel,i,X} = m_{i,X} \cdot \left(\frac{m_{(i-1),X}}{m_{i,X}} - 1 \right)$$

(9)

Where:

$m_{i,x}$ is a certain mass in flight phase i
$m_{(i-1),x}$ is a certain mass in flight phase $(i-1)$

3.2.1 Fuel Consumption due to Fixed Mass

As in Equation **(9)**, the fuel consumption due to fixed mass also has to consider different masses for the different flight phases. The reason why the "fixed" mass could change is due the need to carry the fuel for a later flight phase. In other words, the fixed mass in flight phase $(i-1)$ includes the fuel, which is needed to carry the fixed mass of flight phase i and subsequent phases.

Therefore the calculation of the consumed fuel due to fixed system mass has to start at the end of a flight mission with flight phase $i = 7$. It is agreed, that "...the mass at the *end* of flight phase No. 7 is equal to the (fixed) system mass m_{SYS}..." (**Scholz 1998**). Once the fuel consumption during flight phase 7 is calculated, it has to be added to the system mass and this sum of masses will present the fixed mass for flight phase 6. This procedure will be done for all flight phases and finally all fuel will be added to the overall fuel consumption due to fixed system mass.

Scholz 1998 also presents mass fractions for flight phases $i = 1,2,3,7$. They are shown in the following table.

flight phase	1	2	3	7
m_i/m_{i-1}	1	1	0.995	0.996

Table 1:   Proposed mass fractions m_i/m_{i-1} for aircraft systems

The fuel fractions in flight phases $i = 4,5,6$ can be calculated with an equation also presented by **Scholz 1998**:

$$m_{fuel,i,X} = m_{i,X} \cdot \left(e^{t_i \cdot k_{E,i}} - 1 \right)$$

(10)

Where:
t_i is the duration of flight phase i
$m_{i,x}$ is the mass at the end of flight phase i due to physical cause X
And:

$$k_{E,i} = SFC_i \cdot g \cdot \left(\frac{\cos \gamma_i}{L/D_i} + \sin \gamma_i \right)$$

(11)

Where:
SFC is the thrust specific fuel consumption
g is the earth acceleration
γ is the flight path angle
L/D is the lift to drag ratio

7

3.2.2 Fuel Consumption due to Mechanical Power Off-Takes from the Engines

Scholz 1998 presents an equation for the fuel consumption $m_{fuel,i,P,f}$ due to power off-takes as follows:

$$m_{fuel,i,P,f} = \frac{P_i \cdot k_P \cdot m_{A/C}}{n \cdot T_{T/O}} \cdot \left(e^{t_i \cdot k_{E,i}} - 1 \right)$$

(12)

Where:
P_i is the power off takes in flight phase i
k_P is an average ratio between a thrust specific fuel consumption with and without power off-take and the related power off-take divided by the engine's take-off thrust; **Scholz 1998** gives an average value for k_p with 0.0094 N/W
n is the number of engines
$T_{T/O}$ is the take-off thrust of one engine
t_i is the duration of flight phase i
$k_{E,i}$ is defined as in chapter 3.2.1
$m_{A/C}$ is an average aircraft mass, which could be calculated with the Maximum Take-Off Weight (*MTOW*) and the Maximum Zero Fuel Weight (*MZFW*) as follows:

$$m_{A/C} = \frac{MTOW + MZFW}{2}$$

(13)

For simplicity DOC_{SYS} only considers the power off-takes during flight phases i = 4,5,6. The phases i = 1,2,3,7 could be neglected, as they only present a very short period of a flight mission.

3.3 Direct Maintenance Costs

Scholz 1998 defines DMC for aircraft systems as follows:

$$DMC_{SYS} = \left(MMH_{on} + MMH_{off} \right) \cdot LR + MC$$

(14)

Where:
MMH_{on} are the Maintenance Man Hours On Aircraft
MMH_{off} are the Maintenance Man Hours Off Aircraft
LR is the Labour Rate
MC are the Material Costs

The DMC_{SYS} values used in this thesis represents DMC calculated by an Airbus internal DMC method. The values are representative for typical components as they are installed in today's aircraft. Thus Equation **(14)** will not be further considered. Due to sensitivity, the DMC are shown in non-dimensional form.

3.4 Delay and Cancellation Costs

The extended DOC_{SYS} method, as presented in Equation **(4)**, considers delays and cancellations, because "...operational interruptions cost airlines money." (**Herinckx/Poubeau 2000**). **Scholz 1998** defines Delay and Cancellation Costs for aircraft systems as follows:

$$Delay_{SYS} = \left(D_I \cdot C_I + D_{II} \cdot C_{II} + D_{III} \cdot C_{III} + D_C \cdot C_C \right) \cdot NFY$$

(15)

Where:
D_I is the probability for a delay of up to 29 min.
D_{II} is the probability for a delay between 30 min. and 59 min.

D_{III} is the probability for a delay greater than or equal to 60 min.
D_C is the probability for a cancellation
C_I is the cost of a delay up to 29 min.
C_{II} is the cost of a delay between 30 min. and 59 min.
C_{III} is the cost of a delay of greater than or equal to 60 min.
C_C is the cost of a cancellation
NFY is the number of flights per year

The difficulty in this approach lies in the nature of the input data. It needs a lot of in service experience and thus data from operators to evaluate the probability of Operational Interruptions (OI) due to an aircraft system.

It is also difficult to define an average cost for delays or cancellations, because "…operational interruption costs differ from airline to airline, depending on their respective marketing and operational specifications." (**Herinckx/Poubeau 2000**). Figure 2 and 3 (taken from **Herinckx/Poubeau 2000**) clearly show this difference between the airlines (sources of a questionnaire).

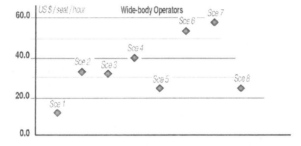

Figure 2: Delay Costs of Wide Body Operators

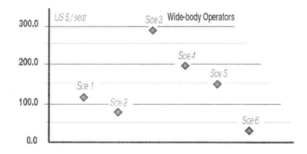

Figure 3: Cancellation Costs of Wide Body Operators

Due to lack of information only the costs due to cancellation will be considered in this thesis. A possible way to evaluate operational interruption costs will be further discussed in chapter 6.2.

3.5 Capital Costs Caused by Spare Parts on Stock

Scholz 1998 defines the costs for spare part holding of aircraft systems as follows:

$$SHC_{SYS} = \frac{SPF \cdot SPR}{RED} \cdot Price \cdot \frac{RQS_{req}}{FS} \cdot r$$ (16)

Where:

SPF is the Spare Part Factor
SPR is the Spare Part Ratio
RED is the average redundancy level
RQS_{req} is the required amount of spare parts
FS is the fleet size
r is interest rate

These costs have not been considered further, and would require procurement information and predicted supplier parts prices to quantify correctly.

4 DecisionPro™ – A Risk Analysis Tool

The algorithms of DOC_{SYS} were incorporated into DecisionPro™ using the available hierarchical tree modelling function. This tool enables construction of interactive tree hierarchies for good visibility of what is happening. It also provides a Monte Carlo Simulation function, which enables stochastic input to be modelled and the impact on a function evaluated. The following simple example should briefly explain the approach of DecisionPro™ and which steps are necessary to use it.

A right-angled triangle is observed with the well-known equation:

$$\sin \alpha = \frac{a}{c}$$ (17)

Equation (17) will be transformed to:

$$c = \frac{a}{\sin \alpha}$$ (18)

4.1 Interactive Tree Hierarchy

The above equation is incorporated into DecisionPro™ using an interactive tree hierarchy approach:

Figure 4: Example of an Interactive Tree Hierarchy in DecisionPro™

The root node on the left hand side defines the triangle hypotenuse c as defined in Equation (18). Once the node is defined, the input nodes, in this case opposite dimension a and substended angle α in radian, appear automatically in the tree structure. In the next step these nodes have to be defined either with a value or with another equation. In this case opposite dimension a has the value 200 without a dimension and α in radian is defined in α in degrees and has the value 45°. Once all nodes are defined, the tree structure can be calculated when the value of each node is specified and is shown in the node boxes. In further steps each input node can easily be modified and the whole tree recalculated.

4.2 Monte Carlo Simulation

DecisionPro™ provides Monte Carlo Simulation to evaluate uncertainties within tree hierarchies and their impact on a function or algorithm. To consider uncertainties, stochastic inputs can be modelled for one or more input nodes. In the case of the above example the angle α could be a stochastic input.

Several stochastic input distributions are available within DecisionPro™:

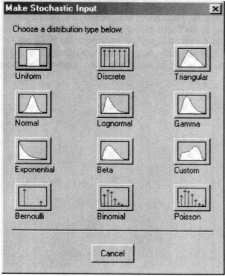

Figure 5: Stochastic Input Distributions within DecisionPro™

For this example the triangular distribution was chosen and the uncertainty was considered by a most likely value and a lower and upper limit for α in degrees.

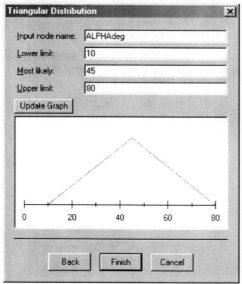

Figure 6: Triangular Distribution for Angle Value

With this distribution DecisionPro™ was able to run a Monte Carlo Simulation by producing random values within the distribution. In other words, DecisionPro™ created several values for α within the boundaries as defined in Figure 6, by respecting the priority of the boundaries as defined. Each random value was then used to evaluate the value of the hypotenuse. The distribution of these values for the numbers of samples specified indicates the likelihood of occurrence for each possible value of c. This is a standard output from DecisionPro™ and is shown in Figure 8 below. Therefore the conditions of the Monte Carlo Simulation were set as follows:

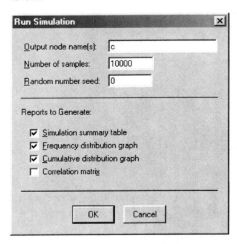

Figure 7:   Monte Carlo Simulation Conditions

The impact on node c, which presents the value of the hypotenuse is being investigated. 10000 random values have been requested by the user and several reports generated.

One of these reports, the Frequency Distribution, shows in how many cases which value was produced. For example approximately 3300 random values produced a value of 250 for c.

Frequency Distribution

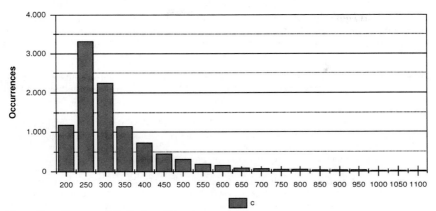

Figure 8: Frequency Distribution of Hypotenuse c

The most useful report is the Cumulative Distribution. This gives an impression of the probability of how an uncertainty affects a result and can be used to directly read the probability of the root node being greater than or less than a specific value. This could be very useful for design purposes where the degree of confidence in a result can be modeled. Of course this could be applied to any engineering problem that can be modeled and requires input parameters that can be estimated with some idea of their uncertainty.

Cumulative Distribution

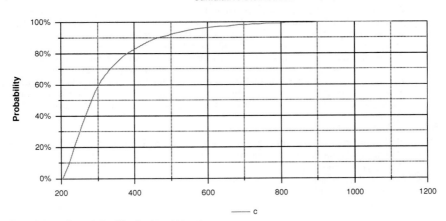

Figure 9: Cumulative Distribution of Hypotenuse c

From the above diagram it can be seen that there is a 60% probability of the hypotenuse c being less than 300.

14

5 Fuel System of a Long Range Aircraft with two Engines

This thesis investigates the DOC of a Long Range Aircraft Fuel System. A configuration with two engines is used. To calculate the DOC a lot of system specific parameters like price, weight, power consumption and maintenance costs are necessary as they are described in chapter 3.

Generally aircraft systems are defined as "...a combination of inter-related items arranged to perform a specific function on an aircraft." (**Scholz 2002**). In modern passenger jets the Fuel System typically provides the following basic functions:

1. Storage of the required fuel volume
2. Supply of fuel to the Engines and the Auxiliary Power Unit (APU)
3. Control of aircraft Centre of Gravity (CG)
4. Provide mass distribution for wing bending moment relief

This chapter gives an overview of the main items of the Fuel System as they are described in the Airbus Maintenance Manuals. It also presents the specific parameters of the items as they are needed for the DOC_{SYS} method. Due to the sensitivity of the parameters only the percentage breakdown of the data is presented, e.g. if a component has a price value of 10%, it means that it costs 10% of the total Fuel System price.

Additionally a few assumptions were made:

- A Fuel System comprises of valves driven by actuators, which need electrical power. However, the power consumption is very low, e.g. compared with the power consumption of pumps. Furthermore many of these valves are only used during abnormal flight conditions, which are not considered in this study. Thus the power consumption of valves is neglected.
- A Fuel System comprises many pipes and connectors. Due to lack of time it was not possible to find the detailed price and weight information of each pipe and connector. Instead particular pipes and connectors were taken and the price and weight of these components were investigated. With these data an average price per weight of the pipework was calculated. A pipework weight breakdown divided by Fuel System subchapters was also available. With the weight breakdown and the average price per weight, average prices of the Fuel System subchapter pipeworks were calculated.
- The pipework causes no maintenance costs. It is assumed, that pipes are once installed and will be scrapped as part of the whole aircraft at the end of its life.

5.1 Fuel Storage

The Fuel Storage consists of several sub functions. These can be defined as follows:
- Content of the required fuel volume
- Drainage of water
- Control of air pressure inside the tanks
- Refuel and Defuel on ground
- Gauging
- Jettison

5.1.1 Tank Layout

Most of the fuel is stored in the wing tanks. These tanks are limited by the wing primary structure. The Wing Box consists of the front & rear spar, a middle spar if applicable, ribs, stringers and the top & bottom skin panels of the wing. Figure 10, Figure 11 and Figure 12 show these structural components. Additionally the wing tanks could be divided in several smaller tanks like collector cells (see chapter 5.2) or outer tanks for wing bending moment relief (see chapter 5.4).

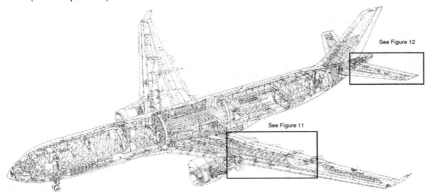

Figure 10: Airbus A330 Cutaway

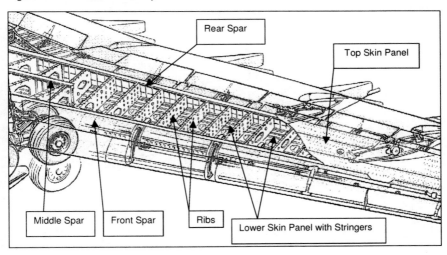

Figure 11: Airbus A330 Wing Box

Today it is common to use a Centre Tank in the fuselage, using the internal bounded volume available in the wing carry through box. This Centre Tank is limited by the Wing Box size and allows short range operators to save costs and weight by using a "No Centre Tank" option. The configuration, which is investigated in this study, will use a "No Centre Tank" option.

A further tank can be placed in the Horizontal Tail Plane (HTP). Similar to the wings, this tail tank is also constrained by the HTP primary structure, as shown in Figure 12. It is typically used to achieve optimum CG positions in cruise for the most efficient aerodynamic performance.

16

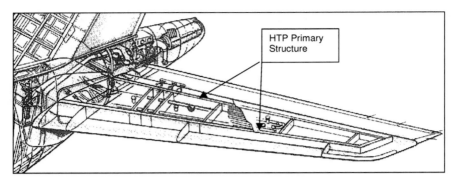

Figure 12: Airbus A330 HTP Box

With this knowledge the Fuel System of the study aircraft consists of five Fuel Tanks. An Inner Tank and an Outer Tank in each wing and a Trim Tank in the HTP. Figure 13 shows this arrangement. Figure 13 also shows three Vent Surge Tanks as they are positioned in each wing tip and the starboard side of the HTP. The vent system will be discussed further in chapter 5.1.3.

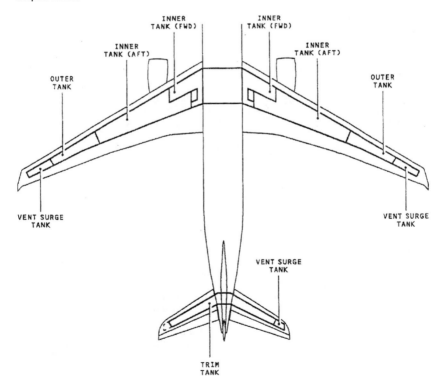

Figure 13: Long Range Aircraft Tank Layout

17

An important consideration for the Tank Layout is the possibility of an Uncontained Engine Rotor Failure (UERF) due to a turbine disc burst. In this case it is assumed that one third of a turbine disc has sufficient energy to penetrate any part of the aircraft. Consequently the inner fuel tanks of the wing could also be damaged. To minimize the fuel loss, each inner tank is divided into a forward and an aft inner tank. Both tanks are connected via a pipe, which can be closed by an Emergency Isolation Valve. The Emergency Isolation Valve includes a motor actuator to open/close the valve. For redundancy a Twin Motor Actuator (TMA) is used, where one motor can open/close the valve if the other motor does not operate. The TMA is installed outside of the tank at the rear spar and is connected via a drive shaft with the valve. This allows maintenance of the TMA without defuelling the tank.

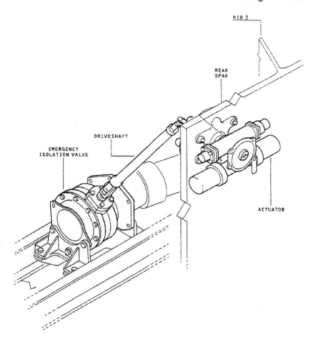

Figure 14: Installation of Emergency Isolation Valve

The two Emergency Isolation Valves are grouped and named as the Emergency Isolation System Valves:

Table 2: Parameters of the Emergency Isolation System Valves

Price [%]	DMC [%]	Weight [%]	Power Consumption [%]
0.961	3.291	1.123	0

5.1.2 Water Drain System

Each tank of the Fuel System has one or more drain valves installed at the lowest point to drain water or remaining fuel out of the tank. The water can occur due to condensation of air moisture inside of the tank. The reason for draining remaining fuel could be due to maintenance.

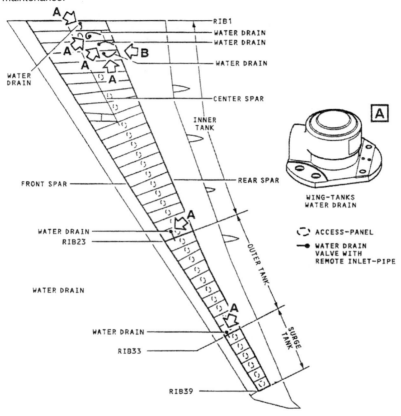

Figure 15: Installation of Water Drain Valves in the Wing Tanks

Figure 15 shows the arrangement of the six Water Drain Valves as they are installed in each wing. There is one valve in each Vent/Surge Tank, each Outer Tank, each Inner Forward Tank and each Collector Cell. There are also two valves in each Inner Aft Tank. The valves are purely mechanical items, which are opened with a special tool by the ground crew.

The Wing Tank related Drain Valves are grouped and named as the Wing Tank Drain System Valves:

Table 3: Parameters of the Wing Tank Drain System Valves

Price [%]	DMC [%]	Weight [%]	Power Consumption [%]
1.492	2.381	0.828	0

In the Trim Tanks are two Direct Drain Valves and in the Trim Vent Surge Tank is one indirect Drain Valve installed as it is shown in Figure 16. The difference between these two valves is the inlet for the fluid to be drained. The direct valve drains the fluid adjacent to the valve. The indirect valve has an inlet connection in the side, which is connected to a pipe that drains the fluid from a short distance from the valve.

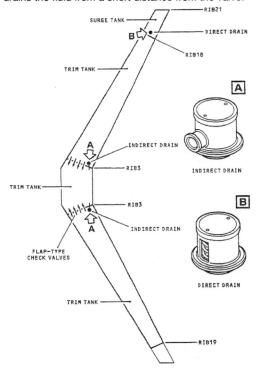

Figure 16: Installation of Water Drain Valves in the Trim Tank

The Trim Tank related Drain Valves are grouped and named as the Trim Tank Drain System Valves:

Table 4: Parameters of the Trim Tank Drain System Valves

Price [%]	DMC [%]	Weight [%]	Power Consumption [%]
0.398	0.099	0.07	0

5.1.3 Venting System

The Venting System has the function to keep the air pressure inside the tanks near to the aircraft's ambient atmosphere pressure. This prevents a large difference between these pressures, which could damage the fuel tank aircraft structure. The vent function is particularly necessary during refuel/defuel operations or during climb and descent. For this reason every fuel tank is connected via venting pipework with the three Vent Surge Tanks, which are located as shown in Figure 13.

5.1.3.1 Wing Tank Venting System

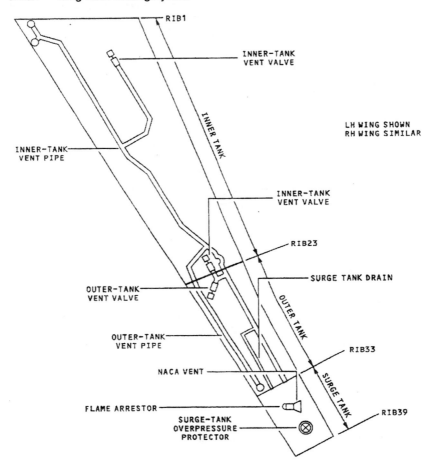

Figure 17: Wing Venting System

The Wing Tank related pipes of the Venting System are grouped and named as the Wing Tank Venting System Pipes:

Table 5: Parameters of the Wing Tank Venting System Pipes

Price [%]	DMC [%]	Weight [%]	Power Consumption [%]
10.384	0	19.912	0

To prevent fuel flow via the vent pipes to the Vent Surge Tanks, the inlets of the pipes are protected with several float valves. The float valves will close the vent pipes if the fuel reaches a certain level. However, it is possible that small amounts of fuel can flow through the vent pipes to the Vent Surge Tank. In this case the Vent Surge Tank has the function of a temporary reservoir.

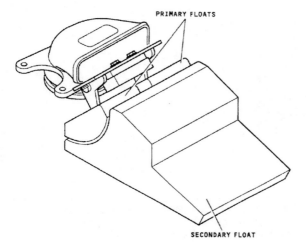

Figure 18: Inner and Outer Vent Valve Rib 23

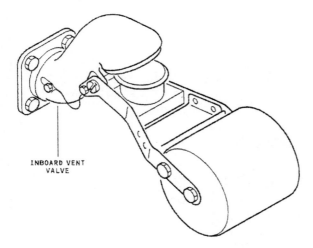

Figure 19: Inner Vent Valve Rib 4

With this arrangement, the air in the Fuel Tanks is able to communicate with the air in the Vent Surge Tanks. The Vent Surge Tank itself communicates with the aircraft's ambient atmosphere via the NACA Flame Arrestor, which lets air flow through it in two directions. The special inner structure of the NACA Flame Arrestor prevents ice formation and ignition of the fuel vapour inside of the Vent Surge Tank in case of a ground fire.

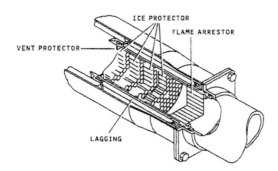

Figure 20: NACA Flame Arrestor

If the airflow through the NACA Flame Arrestor is blocked, an overpressure protector will ensure that the design limits of the Vent Surge Tank will not be exceeded. If the pressure difference between the Vent Surge Tank and the aircraft's ambient atmosphere is more than a specified value, the overpressure protector will break open and release the pressure.

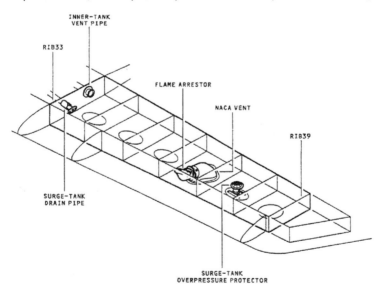

Figure 21: Wing Vent Surge Tank with Overpressure Protector

All the above-described parts of the Wing Venting System are purely mechanical and have no power consumption. The Wing Tanks related float valves, the two NACA Flame Arrestors and the two Overpressure Protectors will be grouped and named as the Wing Tank Venting System Valves.

Table 6: Parameters of the Wing Tank Venting System Valves

Price [%]	DMC [%]	Weight [%]	Power Consumption [%]
4.132	0.088	2.893	0

5.1.3.2 Trim Tank Venting System

Similar to the Wing Tank Venting System the Trim Tank Venting System also consists of several pipes and valves.

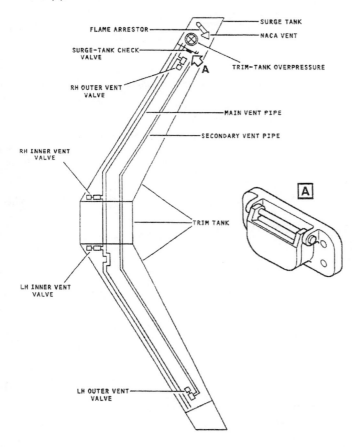

Figure 22: Trim Tank Venting System

The Trim Tank related pipes of the Venting System are grouped and named as the Trim Tank Venting System Pipes:

Table 7: Parameters of the Trim Tank Venting System Pipes

Price [%]	DMC [%]	Weight [%]	Power Consumption [%]
2.314	0	4.437	0

There are two inner and two outer float valves as shown in Figure 22. The valves and pipes provide airflow to the Vent Surge Tank on the starboard side of the HTP. Fuel inside of the Vent Surge Tank can drain via a check valve back to the Trim Tank. The Trim Vent Surge Tank also comprises of a NACA Flame Arrestor and an Overpressure Protector, like the Wing Vent Surge Tank.

The Trim Tank related float valves, the check valve, the NACA Flame Arrestor and the Overpressure Protector will be grouped and named as the Trim Tank Venting System Valves.

Table 8: Parameters of the Trim Tank Venting System Valves

Price [%]	DMC [%]	Weight [%]	Power Consumption [%]
2.28	1.105	1.196	0

5.1.4 Refuel/Defuel System

This system controls the fuel flow into or out of the aircraft on ground.

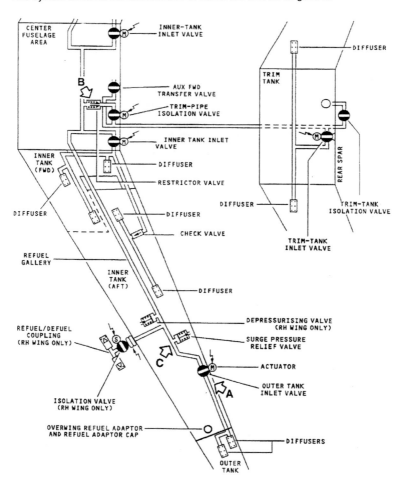

Figure 23: Refuel/Defuel System

The pipes of this system are grouped and named as the Refuel/Defuel Gallery:

Table 9: Parameters of the Refuel/Defuel Gallery

Price [%]	DMC [%]	Weight [%]	Power Consumption [%]
12.242	0	18.664	0

Four different procedures for refuelling or defuelling the aircraft are possible:
- Refuelling:
    o Pressure Refuel
    o Overwing Refuel

- Defuelling:
    o Pressure Defuel
    o Suction Defuel

5.1.4.1   Pressure Refuel

For pressure refuel, one Refuel/Defuel Coupling in the starboard wing connects the Refuel Gallery with the external fuel supply. The Refuel/Defuel Coupling has two valve heads and can therefore be connected to two hoses. An isolation valve inside of the Refuel/Defuel Coupling controls the fuel flow.

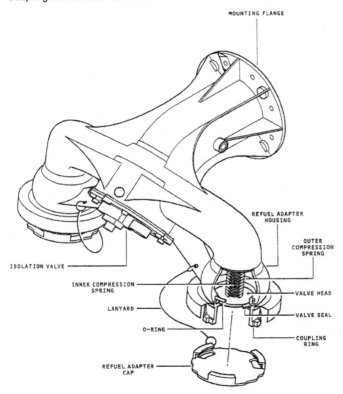

Figure 24:   Refuel/Defuel Coupling

The Refuel Gallery is connected via Inner and Outer Inlet Valves with the several Wing Tanks. A Single Motor Actuator (SMA) drives each inlet valve.

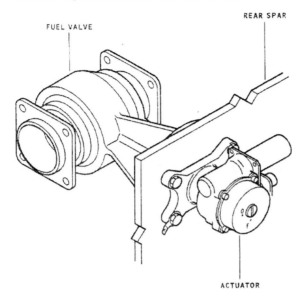

Figure 25: Inlet Valve with Single Motor Actuator

In each tank the Refuel Gallery divides into a number of smaller pipes. At the end of each of these smaller pipes is a refuel diffuser to prevent fuel splash. A check valve prevents fuel flow from the Aft Inner Tank to the Forward Inner Tank.

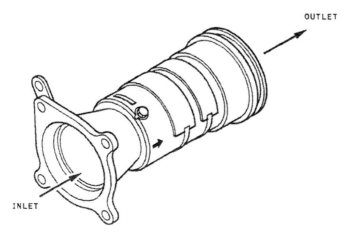

Figure 26: Check Valve Refuel System

The Refuel Gallery is connected via a Restrictor Valve and the Trim Pipe Isolation Valve with the Trim Pipe.

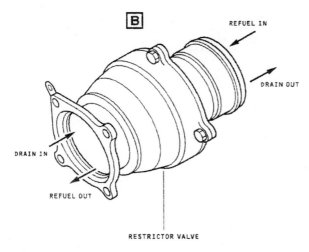

Figure 27: Restrictor Valve

Thus the Trim Pipe will be supplied with fuel and the Trim Tank will be refuelled via the Trim Tank Inlet Valve. The Trim Pipe Isolation Valve and Trim Pipe Inlet Valve belong to the CG Control and will be discussed further in chapter 5.4.

When the refuel operation is completed, the Depressurising Valve in the starboard wing releases the remaining pressure in the Refuel Gallery into the starboard Inner Tank.

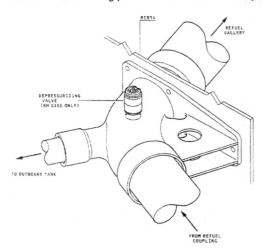

Figure 28: Depressurising Valve

5.1.4.2 Overwing Refuel

For this procedure each Inner Wing Tank consists of an Overwing Refuel Adaptor, which is installed in the top of the tank. The Overwing Refuel Adaptor consists of a cap and the adaptor itself. When the cap is removed, a fuel nozzle could be put into the adaptor to refuel the Inner Tank. Once the Inner Tank is filled a ground transfer with the Engine Feed Pumps is necessary to refuel the remaining tanks. When the ground transfer is done, the Inner Tanks need to again be filled until the required fuel load is achieved. The Engine Feed Pumps will be further described in chapter 5.2.

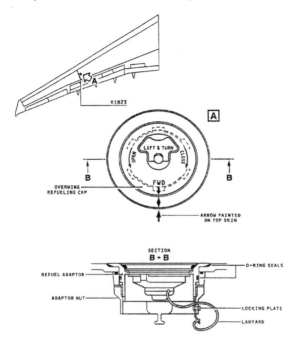

Figure 29: Overwing Refuel Adaptor with Cap

5.1.4.3 Pressure Defuel

In this case the Engine Feed Pumps will push the contained fuel into the Refuel Gallery and from there through the Refuel/Defuel Coupling to an external tank. No extra components are necessary for this procedure.

5.1.4.4 Suction Defuel

In this case tank by tank, fuel will be sucked out. Therefore the related inlet valves will open or close. No extra components are necessary for this procedure.

The two Outer Tank Inlet Valves, the two Inner Tank Inlet Valves, the Refuel/Defuel Coupling, the Restrictor Valve, the two Overwing Refuel Adaptors and Depressurising Valve will be grouped and named as the Refuel/Defuel Valves.

Table 10: Parameters of the Refuel/Defuel Valves

Price [%]	DMC [%]	Weight [%]	Power Consumption [%]
3.643	6.681	3.237	0

5.1.5 Gauging System

The Gauging System can be divided into four subsystems:
- Quantity Indicating System
- Manual Magnetic Indicators
- Tank Level Sensing
- Fuel Temperature Measurement System

5.1.5.1 Quantity Indicating System

This Fuel Quantity Indicating (FQI) System provides information about the quantity of fuel in each tank. Thus several probes are installed in the Inner and Outer Tanks, the two Collector Cells and the Trim Tank.

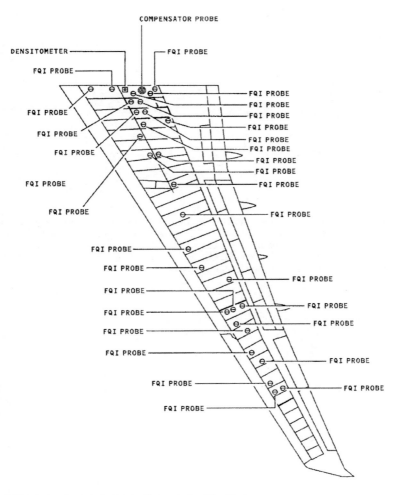

Figure 30: Quantity Indicating Probes in the Wing Tanks

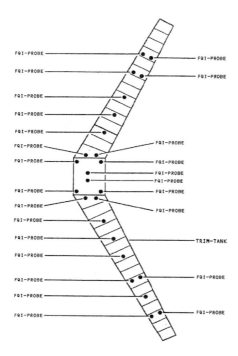

Figure 31: Quantity Indicating Probes in the Trim Tank

Most of these probes are FQI-Probes. Each FQI-Probe comprises two concentric aluminium tubes with different diameters, that are anodised. The capacitance of the probes will change with the level of fuel inside of the probe. The capacitance value is permanently measured by the Fuel Control and Monitoring System (FCMS), which calculates the fuel level with this information. The FCMS will be discussed further in chapter 5.5.

Figure 32: FQI-Probe

One Compensator Probe is installed in each Inner Tank. A Compensator Probe consists of a single tube, which is fixed via two brackets to the tank structure, and two large concentric tubes. The concentric tubes are anodised and the capacitance value is in proportion to the dielectric constant of the fuel.

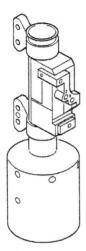

Figure 33: Compensator Probe

One Densitometer is installed in each wing. This component consists of a small resonant system, which will produce certain frequencies at certain fuel densities. In other words, the Densitometer gives permanently a frequency to the FCMS and this value will change if the density of the fuel also changes.

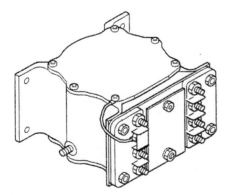

Figure 34: Densitometer

The FCMS uses the information of the FQI-Probes and the Compensator to calculate the volume of the usable fuel. Together with the density value of the Densitometer the FCMS is able to calculate the mass of the usable fuel.

5.1.5.2 Manual Magnetic Indicators

This part of the Gauging System is more a secondary system and can only be used on ground. Four Manual Magnetic Indicators (MMI) are installed in each Inner Wing, two of them are installed in each Outer Wing.

The MMI consists of a Magnetic Level-Indicator (MLI) and a Magnetic Level-Indicator Housing (MLIH).

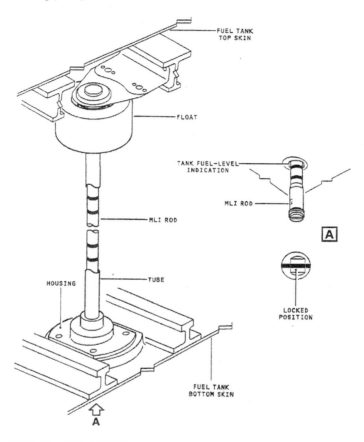

Figure 35: MLI and MLIH

The MLI is a rod with marks to show fuel levels and it is installed in the MLIH. It can be pulled out and pushed back in through an opening in the MLIH. At one end of the rod is a magnet and at the other end is a bayonet-type lock, to clamp the rod inside the MLIH. The MLIH is mounted at the top and the bottom of its related tank. A float assembly around the MLIH can move freely up and down and gives therefore the fuel level. Inside the MLIH is a magnet. The ground crew can now unlock the MLI, extend it slowly to its maximum length and push it slowly back until a magnetic link is felt between the float and the top end of the rod. The reading at the point where the MLI meets the wing skin is the fuel level in that part of the tank.

5.1.5.3   Tank Level Sensing

Several Level Sensors are installed in the tanks to inform the FMCS when a certain critical fuel level is exceeded or falls below a specified minimum level. Therefore the Level Sensors indicate to the FCMS as to whether they are wet or dry.

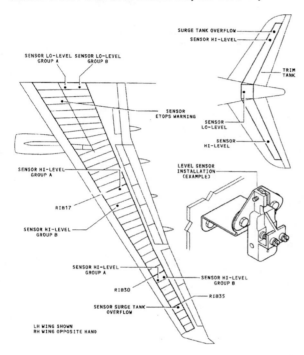

Figure 36:   Tank Level Sensing

There are Hi-Level Sensors for giving warnings if a level will be exceeded and they are installed as follows:
- Two in each Outer Tank
- Two in each Inner Tank
- Two in each Trim Tank

There are Lo-Level Sensors for giving warnings if the level falls below a specified minimum level and they are installed as follows:
- Two in each Inner Tank
- One in the Trim Tank

There are Overflow Sensors installed as follows:
- One in each Wing Surge Tank
- One in the Trim Tank Surge Tank

Each Inner Wing consists of an additional ETOPS (Extended Range Twin Engine Operation) Sensor, which gives warnings if the fuel level falls below a specified minimum level. The minimum level is equal to enough fuel for 180 minutes of flying.

34

5.1.5.4 Fuel Temperature Measurement System

To measure the fuel temperature, four Temperature Sensors are installed in the tanks:
- One in the left Outer Wing Tank
- One in each Collector Cell
- One in the Trim Tank

All the Temperature Sensors are installed at the lowest point of their related tanks. This makes sure, that the sensors are submerged in the fuel for most of the time.

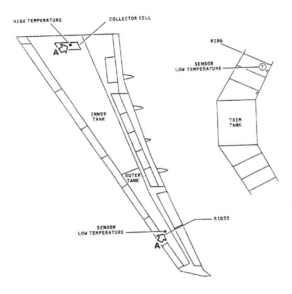

Figure 37: Temperature Sensor Installation

The Inner Wing Tank related FQIs, Compensators, Densitometers, MMIs and Level Sensors are grouped and named as the Inner Wing Tank Probes:

Table 11: Parameters of the Inner Wing Tank Probes

Price [%]	DMC [%]	Weight [%]	Power Consumption [%]
8.92	13.081	3.97	0

The Outer Wing Tank related FQIs, MMIs, Level Sensors, and Temperature Sensors are grouped and named as the Outer Wing Tank Probes:

Table 12: Parameters of the Outer Wing Tank Probes

Price [%]	DMC [%]	Weight [%]	Power Consumption [%]
3.784	6.168	1.153	0

The Trim Tank related FQIs, Level Sensors, and Temperature Sensors are grouped and named as the Trim Tank Probes:

Table 13: Parameters of the Trim Tank Probes

Price [%]	DMC [%]	Weight [%]	Power Consumption [%]
4.094	7.094	1.856	0

5.1.6 Jettison System

Today it is common to provide a Jettison System for aircrafts. With this system the flight crew is able to reduce the aircraft weight in the air to the Maximum Landing Weight (MLW) by jettisoning fuel. This could happen, if a serious failure takes place directly after take-off and therefore an emergency landing is required.

However, a Jettison System is not necessarily required. For example every Airbus aircraft is able to land with the MTOW, but the airline has to do certain checks with the Landing Gear and the aircraft structure afterwards, if the MLW is exceeded. Thus the Jettison System is optional and the operator has to evaluate the probability of a landing with a weight higher than the MLW and the associated maintenance costs versus the operating costs due to a Jettison System.

The configuration in this study will use a "No Jettison System" option and thus typical components for fuel jettison will not be considered again.

5.2 Engine and APU Feed System

5.2.1 Engine Supply

The engines will be supplied with fuel from the Inner Tanks of the wings. Two Main Feed Pumps and one Standby Feed Pump are installed for each engine to maintain the fuel supply. In normal conditions only the Main Feed Pumps will operate and the Standby Feed Pumps will only be used if one of the Main Feed Pumps fail.

All these fuel pumps are installed in canisters. This hollow body is mounted via a mounting flange on the bottom of the tank and connects the fuel pump via a strainer and a slide valve with the tank. The canister could be opened via a bottom access hole to remove the fuel pump. The slide valve is open when the fuel pump is installed and closes automatically when the fuel pump is removed.

The canister has two fuel outlets. One is connected with the engine feed line and thus this outlet provides the real engine fuel supply. The other outlet is connected with a jet pump, whose function will be described later in this chapter. Both outlets are controlled via a check valve. The check valve of the engine feed outlet is installed inside of the canister. The check valve of the jet pump outlet is mounted on the related outlet. Both check valves prevent fuel flow from the engine feed line or the jet pump back into the canister. This design and arrangement of the canister provides maintenance work on the fuel pump without defuelling the related tank.

The canister and thus the pump is connected with a pressure switch, which is placed outside of the tank and gets information of the aircraft's ambient air pressure. This reference pressure is used to control the correct function of the fuel pump. In other words, if the pressure switch recognizes values out of certain limits, the related fuel pump will be set to OFF.

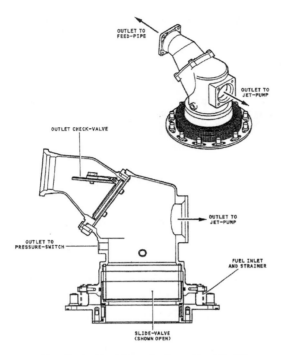

Figure 38: Engine Feed Pump Canister without Fuel Pump

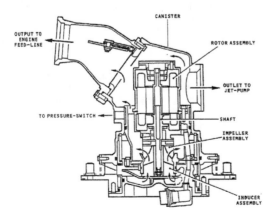

Figure 39: Engine Feed Pump Canister with Fuel Pump

Figure 40 shows the arrangement of the feed pumps as they are installed in the engine feed system.

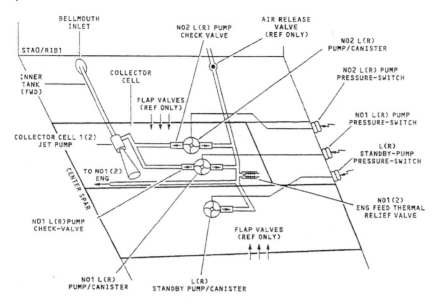

Figure 40: Engine Feed System

The four Main Feed Pumps and the two Standby Feed Pumps have the following parameters:

Table 14: Parameters of the Main Feed Pumps

Price [%]	DMC [%]	Weight [%]	Power Consumption [%]
3.173	5.996	4.349	66.09

Table 15: Parameters of the Standby Feed Pumps

Price [%]	DMC [%]	Weight [%]	Power Consumption [%]
1.586	2.998	2.175	0

Figure 40 shows that both main feed pumps are installed in a Collector Cell. The Collector Cell is a smaller tank within the Inner Wing Tank. The reason for the Collector Cell lies in the design of the fuel pumps. As it is shown in Figure 31 the fuel pump sucks the fuel from the bottom of the related tank. This principle is for redundancy. In other words, if all feed pumps fail, the engine could be fed by gravity. The problem is that the wing tanks are not full during a flight mission. This means, that if the aircraft makes a negative or zero g manoeuvre, the fuel will move up inside of the tank and the tank bottom will become dry. In this case the fuel pumps would run dry and thus the engine would not get any more fuel. To prevent this situation, the Collector Cell is built around the Main Feed Pumps and is permanently full of fuel. The fuel supply for the Collector Cell is mainly done by a jet pump and secondary done by flap valves. The flap valves are installed in a rib at the bottom of the Collector Cell and provide fuel flow into the Collector Cell. These valves are very cheap and light and will be not further considered.

The jet pump is a non-mechanical pump, which consists of a motive flow inlet, a suction inlet, a mixing tube and a diffuser.

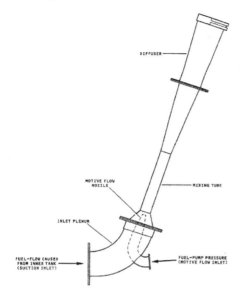

Figure 41: Collector Cell Jet Pump

The motive flow works together with the motive flow nozzle suction inside of the Jet Pump. This transfers fuel from the Inner Tank into the Jet Pump where the motive flow and the suction flow will be mixed and stabilized in the diffuser before going into the Collector Cell. The motive flow is taken from the Main Feed Pumps.

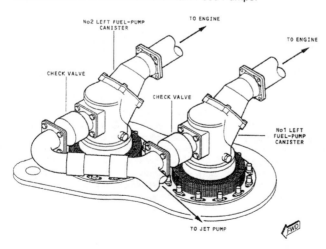

Figure 42: Main Feed Pump Arrangement

The two Collector Cell Jet Pumps have the following parameters:

Table 16: Parameters of the Collector Cell Jet Pumps

Price [%]	DMC [%]	Weight [%]	Power Consumption [%]
0.823	0.284	1.216	0

Additionally a Thermal Relief Valve is installed in the feed line, to prevent too much pressure. It opens if certain limits will be exceeded.

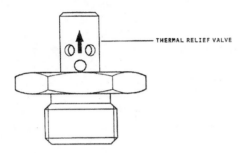

THERMAL RELIEF VALVE

Figure 43: Engine Feed Thermal Relief Valve

An Air Release Valve is installed at the high point of the feed line. It releases trapped air into the Inner Tank.

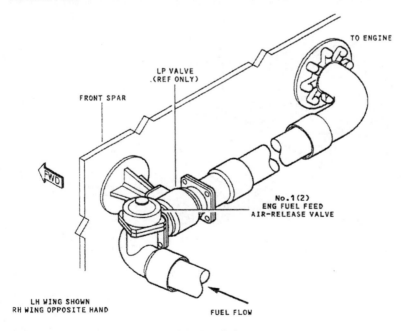

Figure 44: Engine Feed Air Release Valve Installation

40

In the description of the tank layout a possible UERF was mentioned. Also for the engine feed system a certain requirement has to be considered in this case. If the aircraft loses one engine all the stored fuel has to be available for the remaining engine. Thus each engine is connected with the other engine's feed system via the crossfeed system.

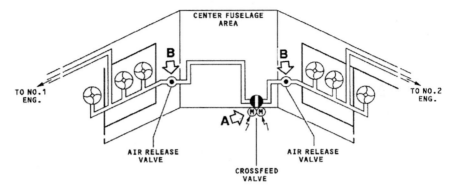

Figure 45: Crossfeed System

Therefore the feed systems of both engines are connected via a pipe, which is normally closed by the Crossfeed Valve. If necessary the Crossfeed Valve will open and all the fuel is available for one engine. The Crossfeed Valve is mounted on the rear spar and driven by a TMA.

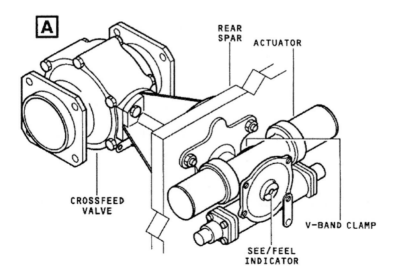

Figure 46: Crossfeed Valve

Two Air Release Valves are installed in the line to the Crossfeed Valve to prevent a build-up of air.

41

If an engine fails, it has to be isolated from the Engine Feed System. Hence a Low Pressure (LP) Valve is installed in the feed line directly before the engine. The LP Valve is mounted on the front spar and is driven by a TMA.

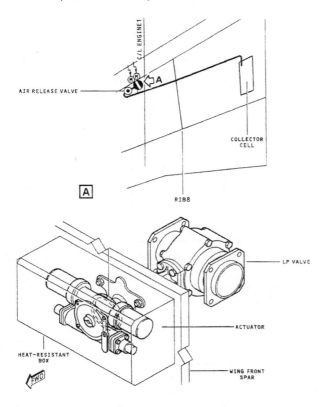

Figure 47: Engine LP Valve

The four Check Valves of the Main Feed Pumps, the two Thermal Relief Valves, the two Air Release Valves in the Engine Feed Line, the Crossfeed Valve, the two Air Release Valves of the Crossfeed Line and the two LP Valves are grouped and named as the Engine Feed System Valves.

Table 17: Parameters of the Engine Feed System Valves

Price [%]	DMC [%]	Weight [%]	Power Consumption [%]
2.671	7.549	1.807	0

The pipes of the Engine Feed Line and the Crossfeed Line are grouped and named as the Engine Feed System Pipes.

Table 18: Parameters of the Engine Feed System Pipes

Price [%]	DMC [%]	Weight [%]	Power Consumption [%]
6.296	0	9.599	0

42

5.2.2 APU Supply

Like the engines the APU also needs fuel supply to operate in all conditions. Therefore a Forward APU Pump is installed in the centre fuselage area between both wings and an Aft APU Pump is installed in the tail cone of the aircraft. Both pumps are installed in canisters.

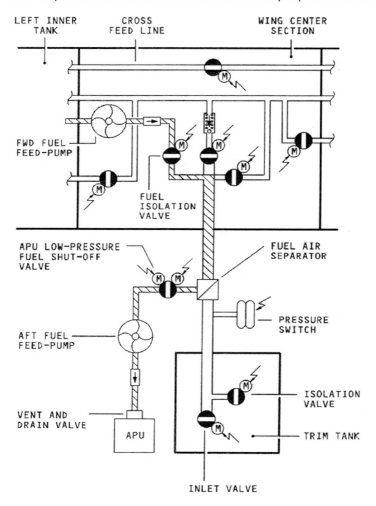

Figure 48: APU Feed System

The Forward APU Pump transfers fuel from the left Inner Wing Tank through the Trim Pipe to the tail cone and the APU when the trim transfer system does not operate. As described in chapter 5.2.1 the Forward APU Pump can also be used to transfer fuel to the engine feed line. The Forward APU Pump could be separated via a Fuel Isolation Valve from the Trim Pipe.

43

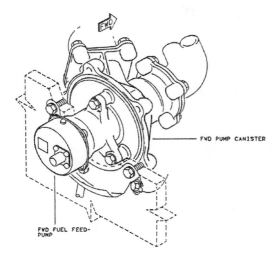

Figure 49: Forward APU Pump

The Aft APU Pump transfers fuel from the Trim Pipe to the APU when the trim transfer system operates. A pressure switch controls this Aft APU Pump. It starts running, when the absolute pressure in the Trim Pipe goes below a specified limit. The Aft APU Pump and the APU can be separated via a Fire Shut-Off Valve from the Trim Pipe. The Fire Shut-Off Valve is driven by a TMA.

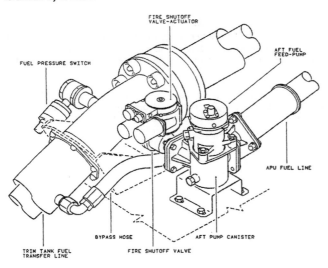

Figure 50: Aft APU Pump Arrangement

A Vent and Drain Valve is installed between the APU and the APU feed line. It is used for the line maintenance to drain or to bleed the APU fuel line.

Table 19: Parameters of the APU Forward Pump

Price [%]	DMC [%]	Weight [%]	Power Consumption [%]
0,762	6,78	0,292	4,533

Table 20: Parameters of the APU Aft Pump

Price [%]	DMC [%]	Weight [%]	Power Consumption [%]
2,196	1,637	0,842	7,556

The APU Fuel Isolation Valve and the APU Fire Shut-Off Valve are grouped and named as the APU Feed System Valves.

Table 21: Parameters of the APU Feed System Valves

Price [%]	DMC [%]	Weight [%]	Power Consumption [%]
1,323	0,612	0,412	0

The pipes for the APU fuel supply are grouped and named as the APU Feed System Pipes.

Table 22: Parameters of the APU Feed System Pipes

Price [%]	DMC [%]	Weight [%]	Power Consumption [%]
0,35	0	0,533	0

5.3   CG Control

Due to variable passenger and cargo distribution the CG position of the aircraft will be different for each flight mission. Due to the fact, that the engines will burn most of the stored fuel, the CG position will also move during a flight mission. The lift force produces a penalising moment to the moving CG. In small aircrafts the HTP is used for producing a down force and thus a contra moment to the lift force. For this principle the angle of attack of the HTP changes and this will cause increased drag and fuel consumption.

For larger aircraft it could be more efficient to use a Trim Tank in the HTP where fuel is moved between the wing and HTP related tanks to change the position of the CG during the flight. The penalty of this principle lies in a more complex fuel system due to extra equipment. As it is described in chapter 5.1 the configuration investigated in this thesis should consist of a Trim Tank.

To transfer fuel from the Trim Tank to the Wing Tanks (Forward Transfer) or from the Wing Tanks to the Trim Tanks (Aft Transfer) several components are necessary. In normal flight conditions one small Aft Transfer and several Forward Transfers will be done.

5.3.1 Aft Transfer

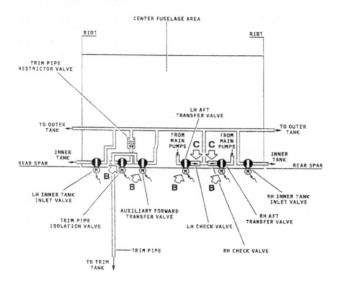

Figure 51: Trim Transfer System Centre Fuselage Area

Via two Aft Transfer Valves, the Engine Feed Lines can supply fuel to the Refuel/Defuel Gallery. The Aft Transfer Valves are mounted on the rear spar and driven by a SMA. Each Aft Transfer Valve is connected to a Check Valve, which prevents fuel flow from the Refuel/Defuel Gallery back into the Engine Feed Lines.

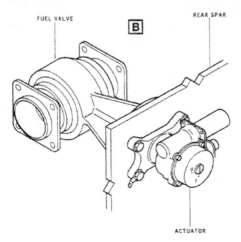

Figure 52: Fuel Valve with SMA

46

The opened Trim Pipe Isolation Valve connects the Refuel/Defuel Gallery with the Trim Pipe. Like the Aft Transfer Valves this valve is mounted on the rear spar and driven by a SMA. The Auxiliary Forward Transfer Valve and all Outer Tank and Inner Tank Inlet Valves are closed during this procedure. Thus the fuel from the Refuel/Defuel Gallery can flow through a Trim Pipe Restrictor Valve and the opened Trim Pipe Isolation Valve into the Trim Pipe, which is connected with the Trim Tank in the HTP.

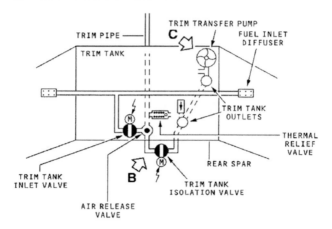

Figure 53: Trim Transfer System in the HTP

The Trim Pipe is connected via a Trim Tank Inlet Valve with the Trim Tank. The Trim Tank Inlet Valve is mounted on the bottom and driven by a SMA. During an Aft Transfer this valve is opened and therefore the fuel from the Trim Pipe can flow into the Trim Tank.

5.3.2 Forward Transfer

The Trim Tank Transfer Pump transfers fuel via Trim Tank Outlets and the opened Trim Tank Isolation Valve from the Trim Tank into the Trim Pipe. A SMA drives the Trim Tank Isolation Valve. One outlet consists of a check valve and an Air Release Valve. In addition a Thermal Relief Valve is installed at the Trim Tank inlet and outlet pipe.

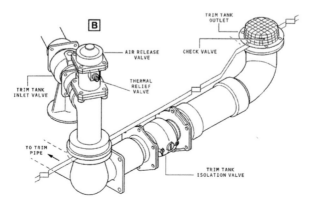

Figure 54: Trim Tank Inlet Valve & Trim Tank Isolation Valve

The Trim Pipe is connected via the opened Auxiliary Forward Valve with the Refuel/Defuel Gallery. The Trim Pipe Isolation Valve and both Aft Transfer Valves are closed during this procedure. The Inner Tank Inlet Valves are opened during this procedure and thus the fuel can flow from the Refuel/Defuel Gallery into the Inner Wing Tanks.

Two small jet pumps are connected with the Trim Tank Transfer Pump. These Trim Tank Scavenge Jet Pumps mix the fuel inside of the Trim Tank to prevent big bubbles of water.

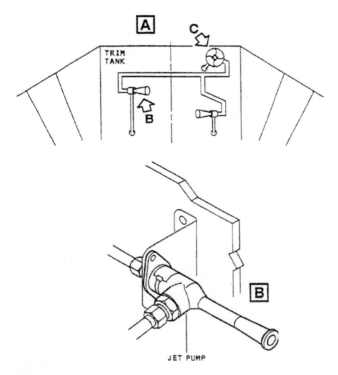

Figure 55: Trim Tank Scavenge Jet Pump

Table 23: Parameters of the Trim Tank Transfer Pump

Price [%]	DMC [%]	Weight [%]	Power Consumption [%]
2,305	2,767	1,069	17,469

Table 24: Parameters of the Trim Tank Scavenge Jet Pumps

Price [%]	DMC [%]	Weight [%]	Power Consumption [%]
0,243	0,033	0,337	0

Table 25: Parameters of the Trim Tank Transfer System Valves

Price [%]	DMC [%]	Weight [%]	Power Consumption [%]
4,113	9,492	1,379	0

Table 26: Parameters of the Trim Tank Transfer System Pipes

Price [%]	DMC [%]	Weight [%]	Power Consumption [%]
7,867	0	11,994	0

5.4 Wing Bending Moment Relief

Due to lift distribution the wings will produce a bending moment at the wing root. The wing structure has to be designed to account for the bending moment. To relieve the bending moment and thus the wing structure Outer Wing Tanks are installed. The fuel inside of these tanks produces a down force and thus a contra moment to the lift related moment. This fuel is kept for as long as possible in the Outer Wing Tanks to relieve the bending moment.

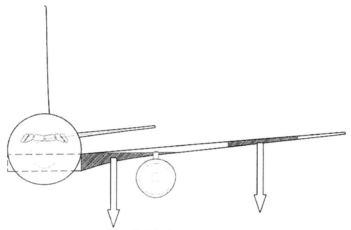

Figure 56: Down Forces by Fuel Tanks

When the engines burn fuel during a flight mission they take the fuel from the Inner Wing Tanks as described in chapter 5.2.1. To control the changing CG position due to the fuel burn, fuel from the Trim Tank will be transferred in intervals to the Inner Wing Tanks as described in chapter 5.3.2.

When the Trim Tank is empty, the remaining fuel inside of the Wing Tanks will burn to a certain level. Then the fuel from the Outer Wing Tanks will be transferred into the Inner Wing Tanks. This transfer will also be done in intervals.

For this reason each Outer Wing Tank consists of an Intertank Transfer Valve. When this valve opens, the fuel from the Outer Wing Tank is transferred by gravity into the Inner Wing Tank. The Intertank Transfer Valve is mounted on a rib and driven via a driveshaft by a SMA.

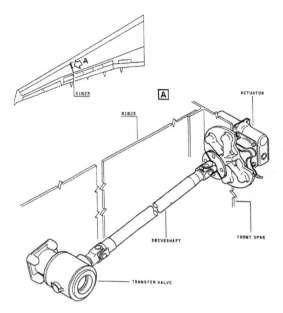

Figure 57: Intertank Transfer Valve

Table 27: Parameters of the Intertank Transfer System Valves

Price [%]	DMC [%]	Weight [%]	Power Consumption [%]
0,888	2,944	0,635	0

5.5 Fuel Control and Monitoring System

To control the components and functions of such a Fuel System a Fuel Control and Monitoring System (FCMS) is necessary. The major parts of the FCMS are two Fuel Control and Monitoring Computers (FCMC).

The FCMC receive various data from the probes and status information from valves and pumps. With this information the FCMC are able to calculate certain values and make the applicable decisions for:

- Fuel Quantity and Fuel Temperature Measurement
- Trim Transfer
- Intertank Transfer
- Automatic Pressure Refuel
- Manual Refuel/Defuel and Ground Transfer

Only one FCMC controls the Fuel System at a time. The other one continuously monitors the responsible values and gets control if the first FCMC gives unsatisfactory data.

The FCMC have the following specific parameters:

Table 28: Parameters of the FCMC

Price [%]	DMC [%]	Weight [%]	Power Consumption [%]
10,762	21,994	4,023	4,352

5.6 Fuel System Parameter Overview

The collected data tabulated above was used to create diagrams regarding the inputs of DOC_{SYS}. To simplify the overviews the components are grouped as pumps, valves, gauging, pipes and the FCMS.

5.6.1 Fuel System Price

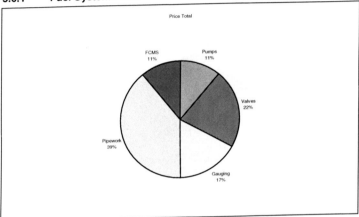

Figure 58: Price Breakdown by Component Groups

The pipes contribute the biggest fraction of the Fuel System price with 39%. The valves with 22% and the gauging system with 17% follow them. The prices of the pumps and the two FCMC are equal with 11%.

5.6.2 Fuel System DMC

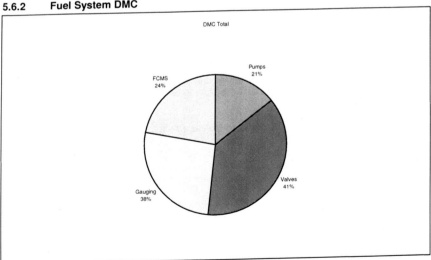

Figure 59: DMC Breakdown by Component Groups

The valves contribute the biggest fraction of the Fuel System DMC with 41%. The gauging with 38% and the FCMS with 24% follow them. The pumps cause the smallest DMC fraction with 21%.

5.6.3   Fuel System Weight

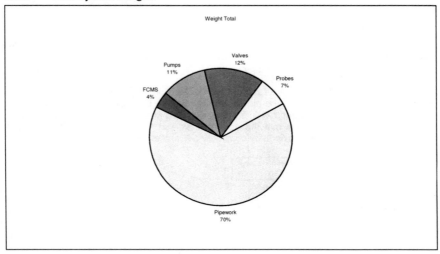

Figure 60:   Weight Breakdown by Component Groups

The pipes contribute by far the biggest fraction of the Fuel System weight with 70%. The weight fractions of the valves and pumps are equal with 11% and 12%. The Probes and the FCMS contribute small fractions with 7% and 4%.

6 Direct Operating Costs of a Long Range Aircraft Fuel System

The data as described in chapter 5 was used to evaluate DOC of a Long Range Aircraft Fuel System Architecture. Thus the algorithms of DOC_{SYS} were programmed with DecisionPro™.

6.1 Application of DOC$_{SYS}$ for a Fuel System with DecisionPro™

To validate the combination of the DOC_{SYS} method and DecisionPro™ a first test run was made in accordance with the commercially available program DOC_{SYS}. The inputs for this test run do not present a Fuel System of a particular aircraft type.

As explained in chapter 4 the tree structure within DecisionPro™ starts with the root node, which presents the final result of the model. In this case the DOC of a Fuel System per year were calculated. To provide good visibility and minimize space requirements the following tree structures are only shown with the node names and the node values. Thus the tree hierarchy has the following root node by considering Equation **(3)**:

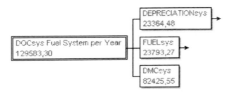

Figure 61: Root Node of Fuel System DOC per Year

6.1.1 Depreciation

The depreciation node is defined as in Equation **(5)** with a price, a residual and a depreciation period. The price was 389408 $. The residual is calculated by a factor of 0.1 of the price and the depreciation period is assumed with 15 years. With these information the branch structure of the depreciation is as follows:

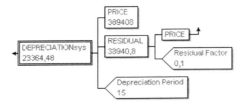

Figure 62: Depreciation Branch

6.1.2   Fuel Burn

The fuel node is defined as shown in Equation **(8)**. The NFY is assumed with 700 and the fuel price is assumed with 0,3419 $/kg:

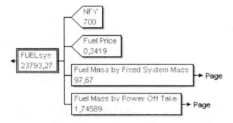

Figure 63:   Fuel Costs Branch

In a next step the fuel masses were defined.

6.1.2.1   Fuel Mass due to Fixed System Mass

The fuel mass due to fixed system mass node is defined as it is described in chapter 3.2.1. Thus the seven flight phases and their contribution to the fuel consumption were considered:

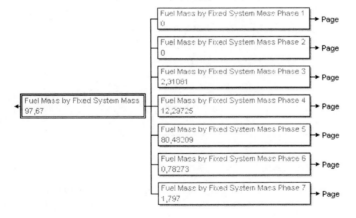

Figure 64:   Fuel Mass due to Fixed System Mass Branch

6.1.2.1.1 Flight Phase 7

As also described in chapter 3.2.2 the calculation has to start at the end of a flight mission by only considering the system mass. Thus the node of flight phase 7 is defined as follows:

Figure 65: Fuel Mass due to Fixed System Mass Phase 7 Branch

The fuel consumption for this flight phase is defined as in Equation **(9)**. The mass fraction was taken from Table 1 and the fixed system mass was assumed with 447.554 kg.

6.1.2.1.2 Flight Phase 6

Flight phase 6 was considered with the following branch:

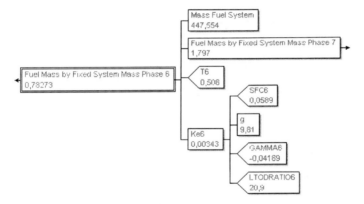

Figure 66: Fuel Mass due to Fixed System Mass Phase 6 Branch

The fixed system mass in this flight phase is the sum of the system mass and the required fuel mass for flight phase 7. The fuel consumption for flight phase 6 was defined as in Equation **(10)**. The following inputs were assumed as follows:

t_6 = 0.508 hr

The factor $k_{E,6}$ was defined as in Equation **(11)**, with the following assumed inputs:

SFC_6 = 0,0589 kg/(hr*N)
g = 9.81 m/s2
γ = -0.04189 rad
L/D = 20.9

6.1.2.1.3 Flight Phase 5

Flight phase 5 was considered with the following branch:

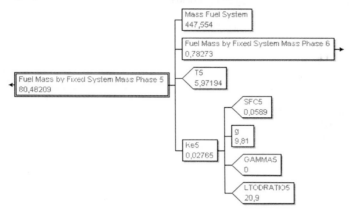

Figure 67:   Fuel Mass due to Fixed System Mass Phase 5 Branch

The fixed system mass in this flight phase is the sum of the system mass and the required fuel mass for flight phase 6. The fuel consumption for flight phase 5 was defined as in Equation **(10)**. The following inputs were assumed as follows:

t_5 =   5.967 hr

The factor $k_{E,5}$ was defined as in Equation **(11)**, with the following assumed inputs:

SFC_5 =   0,0589 kg/(hr*N)
g =   9.81 m/s2
γ =   0 rad
L/D =   20.9

6.1.2.1.4 Flight Phase 4

Flight phase 4 was considered with the following branch:

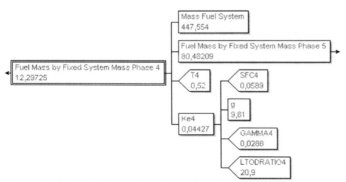

Figure 68:   Fuel Mass due to Fixed System Mass Phase 4 Branch

The fixed system mass in this flight phase is the sum of the system mass and the required fuel mass for flight phase 5. The fuel consumption for flight phase 4 was defined as in Equation **(10)**. The following inputs were assumed as follows:

t_4 =   0.52 hr

The factor $k_{E,4}$ was defined as in Equation **(11)**, with the following assumed inputs:

SFC_5 =   0,0589 kg/(hr*N)
g =   9.81 m/s2
γ =   0.0288 rad
L/D =   20.9

6.1.2.1.5 Flight Phase 3

Flight phase 3 was considered with the following branch:

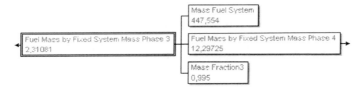

Figure 69: Fuel Mass due to Fixed System Mass Phase 3 Branch

The fixed system mass in this flight phase is the sum of the system mass and the required fuel mass for flight phase 4. The fuel consumption for flight phase 3 was defined as in Equation **(9)**. The mass fraction was taken from Table 1.

6.1.2.1.6 Flight Phase 2

Flight phase 2 was considered with the following branch:

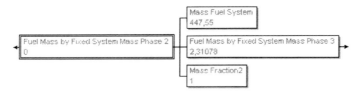

Figure 70: Fuel Mass due to Fixed System Mass Phase 2 Branch

The fixed system mass in this flight phase is the sum of the system mass and the required fuel mass for flight phase 3. The fuel consumption for flight phase 2 was defined as in Equation **(9)**. The mass fraction was taken from Table 1.

6.1.2.1.7 Flight Phase 1

Flight phase 2 was considered with the following branch:

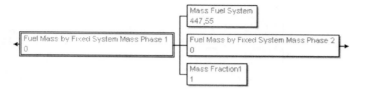

Figure 71: Fuel Mass due to Fixed System Mass Phase 1 Branch

The fixed system mass in this flight phase is the sum of the system mass and the required fuel mass for flight phase 2. The fuel consumption for flight phase 1 was defined as in Equation **(9)**. The mass fraction was taken from Table 1.

6.1.2.2 Fuel Mass due to Power Off-Take

The fuel mass due to fixed system mass node is defined as it is described in chapter 3.2.2. Thus the flight phases 4 to 6 were considered in this branch:

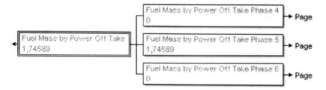

Figure 72: Fuel Mass due to Power Off-Take Branch

No power consumption was considered for flight phases 4 and 6.

6.1.2.2.1 Flight Phase 5

Flight phase 5 was considered with the following branch:

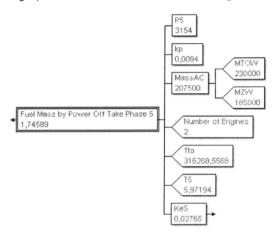

Figure 73: Fuel Mass due to Power Off-Take Phase 4 Branch

The fuel consumption for flight phase 5 was defined as in Equation **(12)**. The inputs were assumed as follows:

P_5	=	3154 W
k_P	=	0.0094
n	=	2
$T_{T/O}$	=	316268.5568 N

The average aircraft mass $m_{A/C}$ was defined as in Equation **(13)**. The inputs were assumed as follows:

$MTOW$	=	230000 kg
MZW	=	185000 kg

The factor $k_{E,5}$ was defined and calculated as in chapter 6.1.2.1.3.

6.1.3 DOC Comparison with the Program DOC$_{SYS}$

The above input parameters were also used to make a run with the commercial available program DOC$_{SYS}$ written by Dieter Scholz to check the above tree structure. The program DOC$_{SYS}$ produced among other things the following outputs:

```
BETRIEBSKOSTEN DES FLUGZEUG(TEIL)SYSTEMS
=========================================

                              US$/Flugzeug/Jahr   Anteil in %
--------------------------------------------------------------------

DOCsys                     =      129782.86      100.00

--------------------------------------------------------------------

Abschreibungskosten        =       23364.48       18.00

Kosten durch Wartung und
Instandhaltung             =       82425.55       63.51

Kraftstoffkosten durch den
Transport von fixen Massen =       23562.94       18.16

Kraftstoffkosten durch
Wellenleistungsentnahme    =         429.89        0.33
```

Where:

Abschreibungskosten	=	Depreciation
Kosten durch Wartung und Instandhaltung	=	Maintenance Costs
Kraftstoffkosten durch den Transport von fixen Massen	=	Fuel Costs due to fixed Mass
Kraftstoffkosten durch Wellenleistungsentnahme	=	Fuel Costs due to Power Off-Takes

The Fuel System DOC with the DOC$_{SYS}$ Program were calculated with 129782.86 \$ and as shown in Figure 61 using DecisionPro^{TM} with 129583.30 \$, using the same input data. Therefore the calculation with DecisionPro^{TM} produced a relative error of:

$$\left| \frac{129583.30\$ - 129782.86\$}{129782.86\$} \right| \cdot 100 = 0.154\% \tag{19}$$

This error occurred due to evaluation of the $k_{e,i}$ factor in 1/hr and not in 1/s and the fact that the DOC$_{SYS}$ program also considers the fuel required to carry the fuel to provide the power required by the system during a mission.

Nevertheless the error was seen as very small and therefore the above tree structure was used for further calculations.

6.2   Cancellation Cost Evaluation with Monte Carlo Simulation

As mentioned in chapter 2.5 it is difficult to evaluate DOC caused by OI. To consider the uncertainties within Equation (15), the Monte Carlo Simulation of DecisionPro™ was used. Only costs due to cancellations were investigated. Thus the tree structure from Figure 61 was taken with the same inputs. Additionally a branch for cancellation costs was attached to the root node.

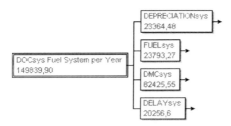

Figure 74:   Tree Structure with OI Cost Node

The node of the cancellation costs was defined as follows:

Figure 75:   Cancellation Cost Node

Where:

NFY	is the Number of Flights per Year
PAXNUM	is the number of passengers
Cancellation Cost	is the cancellation cost per passenger
Cancellation Probability	is the cancellation probability due to the system

The NFY is taken from chapter 6.1 and the number of passengers is assumed with 300.

A stochastic input was made for the cancellation cost based on the data presented in Figure 3. A triangular distribution was chosen with a most likely value of 140 $, a lower limit of 25 $ and an upper limit of 290 $.

A second stochastic input was made for the cancellation probability. Only a statistic about OI rates of a Long Range Aircraft in general was available. However, this statistic was used for this calculation. Figure 77 shows the probability in percent of an OI due to the Fuel System of a Long Range Aircraft for one year. This Figure also shows, that these values vary significantly from month to month.

Therefore a triangular distribution was chosen for the cancellation probability node. The upper limit was 0.001021 and the lower limit was 0.000283 taken from Figure 77. The most likely value was the average OI rate also taken from Figure 77, giving a value of 0.000689.

OI Rates of one year due to the Fuel System

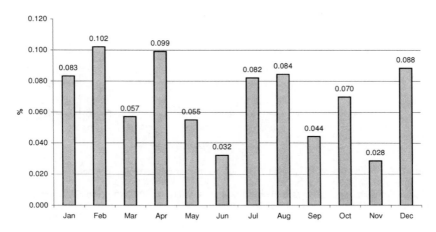

Figure 76: Variation of OI Rates of a Long Range Aircraft due to the Fuel System

With the stochastic inputs for the cancellation costs and the cancellation probability a Monte Carlo Simulation with 10000 samples was done, regarding the DOC of a Fuel System in $ per year. The results were the following:

Frequency Distribution

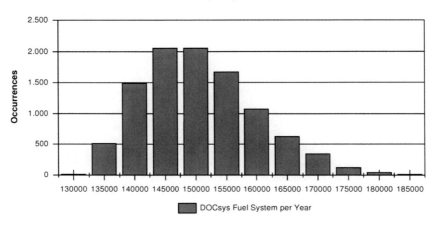

Figure 77: Occurrences of Fuel System DOC per year due to cancellation uncertainty

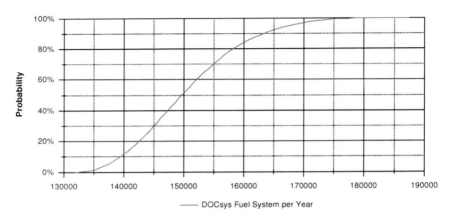

Figure 78: Probability of Fuel System DOC per year due to cancellation uncertainty

The above distribution shows that there is a probability of 12 % that the DOC will be less than 140000 $ and a probability of 84 % that the DOC will be less than 160000 $. In other words, with the chosen distributions of uncertainties the chance is greater that the DOC will be higher than the most likely value of approximately 149840 $ per year.

6.3   Fuel System DOC by DOC_{SYS} Fractions

The data of the components described in chapter 5 were used to calculate the DOC of a whole Fuel System of a today's Long Range Aircraft. A complete flight mission was observed.

The TLAR such as duration of flight phases, flight path angle, SFC, MTOW, etc. were taken from an Airbus internal aircraft configuration tool and are representative for a Long Range Aircraft with two engines. The fuel price was taken from **Hume 2004**.

Similar to chapter 6.1 the algorithms of DOC_{SYS} were also programmed in DecisionPro™. In opposite to chapter 6.1 the DOC were calculated in $ per Flight Hour (FH). Therefore the root node is defined as follows:

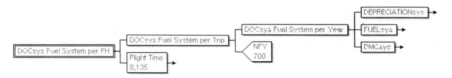

Figure 79:   Root Node for DOC per FH

The DOC_{SYS} method provides the DOC in $ per year. Thus the above branch divides this DOC value with the NFY and further with the Flight Time to achieve the DOC per FH. The branches for depreciation, fuel burn and DMC were defined and structured as in chapter 6.1.

Finally the calculation was done and produced the following breakdown by the fractions of DOC_{SYS}:

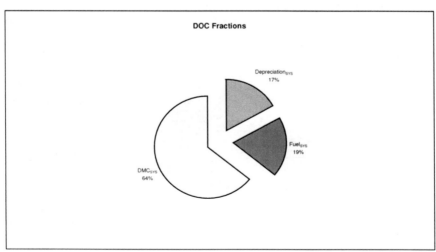

Figure 80:   Fuel System DOC by DOC_{SYS} Fractions

Nearly two thirds of the DOC of the investigated Fuel System Architecture are caused by the DMC. The Fuel Burn and the Depreciation are nearly equal with 19% and 17%.

This calculation also gave an impression of the Fuel Burn fractions:

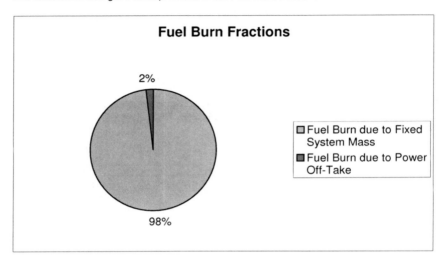

Figure 81: Fuel Burn Fractions

The above diagram shows that the Fuel Burn due to Power Off-Take only contributes a very small fraction to the total Fuel Burn.

6.4   Fuel System DOC by Components

In a next step the influence of the several components to the DOC was evaluated. Therefore it was necessary to calculate the DOC of each component.

Thus the root node was defined as follows:

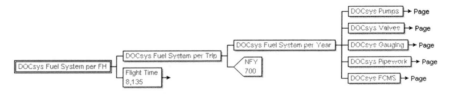

Figure 82: Root Node for DOC$_{SYS}$ per Components

In this case the DOC per year were defined with the DOC of the component groups. As shown in the next five pictures each component group was defined by the related components or subgroups.

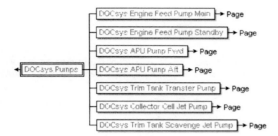

Figure 83: Pumps DOC Branch

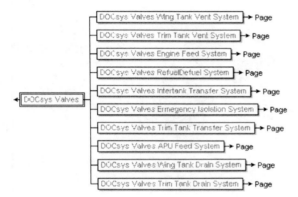

Figure 84: Valves DOC Branch

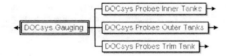

Figure 85: Gauging DOC Branch

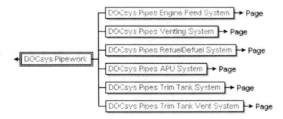

Figure 86: Pipework DOC Branch

Figure 87: FCMS DOC Branch

Afterwards a tree structure for each DOC Component Node was defined and created as for the whole Fuel System in chapter 6.1 and 6.3. The component specific parameters and the TLAR as in chapter 6.3 were considered.

Finally the tree structure was calculated and produced the following results:

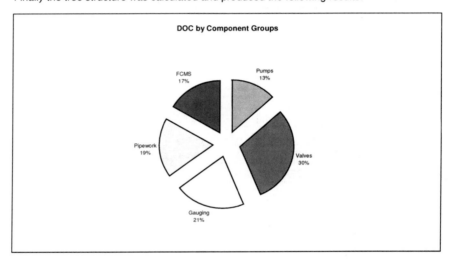

Figure 88: Fuel System DOC by Component Groups

The above diagram shows the DOC breakdown by Component Groups. Within this breakdown the valves take the biggest amount of the DOC with 30%. Equal are the Gauging System and the Pipework with 21% and 19%. The two FCMC contribute with 17% and the pumps with 13% affect the rest of the DOC.

More interesting is the impact on the DOC by the components itself:

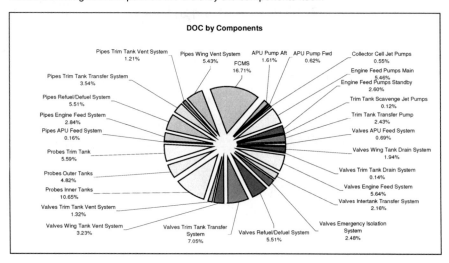

Figure 89: Fuel System DOC by Components

The above diagram shows the contributions of the Fuel System components to the DOC. For simplicity only the top five main drivers will be further discussed. They are:

- The two FCMC with 16.71%
- The Inner Tank Probes with 10.65%
- The Trim Tank Transfer System Valves with 7.05%
- The Engine Feed System Valves with 5.64%
- The Trim Tank Probes with 5.59%

The DOC fractions of these main drivers were calculated as follows:

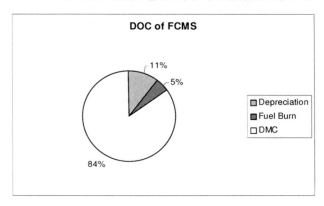

Figure 90: DOC Fractions of FCMS

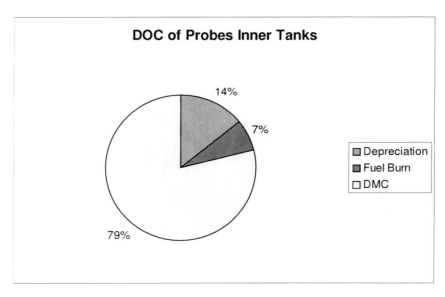

Figure 91: DOC Fractions of Inner Tank Probes

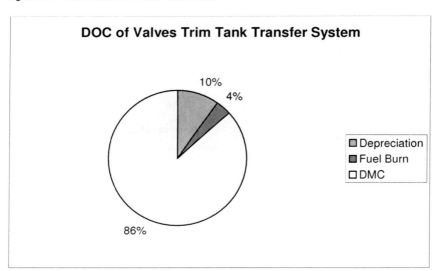

Figure 92: DOC Fractions of Trim Tank Transfer System Valves

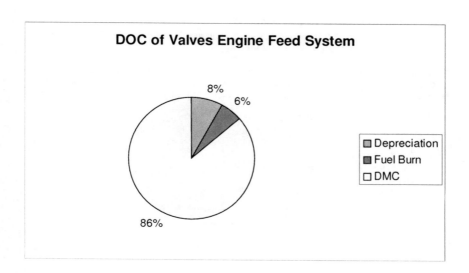

Figure 93: DOC Fractions of Engine Feed System Valves

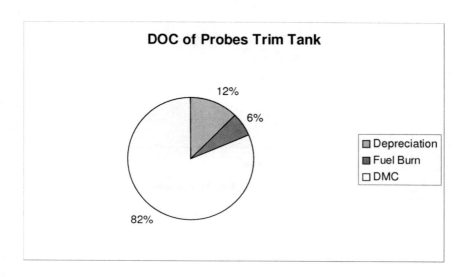

Figure 94: DOC Fractions of Trim Tank Probes

Similar to Figure 81, Figure 91 to Figure 95 demonstrate that the DMC have the greatest impact on the DOC.

6.5 Fuel System DOC by Functions

The results of chapter 6.3 were also used to evaluate the DOC of the Fuel System Main Functions as they are described in chapter 5.

To achieve this overview the several components needed to be grouped in relation to the functions. The following subchapters explain which components were contributed to the several functions and what their impact on the DOC per Function is.

6.5.1 DOC of APU Fuel Supply

The components needed for this function are:

- The Aft APU Pump
- The Forward APU Pump
- The APU Feed System Valves
- The APU Feed System Pipes

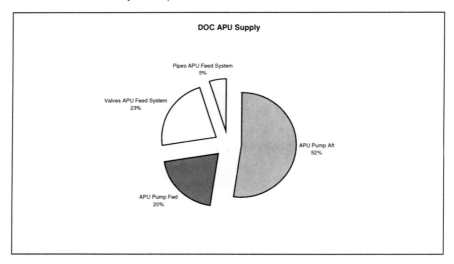

Figure 95: DOC Fractions of APU Fuel Supply

The biggest fraction within the DOC of the APU Fuel Supply is the Aft APU Pump. This occurs due to high DMC of this component.

6.5.2   DOC of Engine Fuel Supply

The components needed for this function are:

- The Main Engine Feed Pumps
- The Standby Engine Feed Pumps
- The Collector Cell Jet Pumps
- The Engine Feed System Valves
- The Engine Feed System Pipes

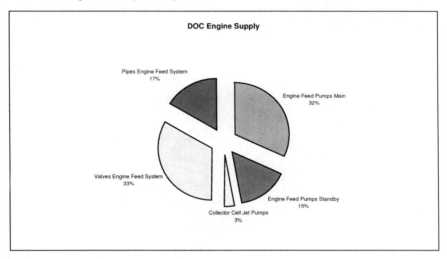

Figure 96:   DOC Fractions of Engine Fuel Supply

The DOC main drivers of the Engine Fuel Supply are the Engine Feed Pumps and the Engine Feed System Valves.

6.5.3   DOC of Wing Bending Moment Relief

This function only comprises the two Intertank Transfer Valves. Thus the DOC fraction of this function is very small.

6.5.4 DOC of Fuel Storage

The components needed for this function are:

- The Emergency Isolation System Valves
- The Refuel/Defuel System Valves
- The Wing Tank Vent System Valves
- The Wing Tank Drain System Valves
- The Inner Tank Probes
- The Outer Tank Probes
- The Refuel/Defuel Gallery
- The Wing Tank Vent System Pipes

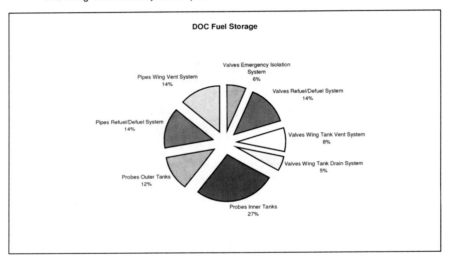

Figure 97: DOC Fractions of Fuel Storage

6.5.5 DOC of the CG Control

The components needed for this function are:

- The Trim Tank Transfer Pump
- The Trim Tank Scavenge Jet Pumps
- The Trim Tank Transfer System Valves
- The Trim Tank Vent System Valves
- The Trim Tank Drain System Valves
- The Trim Tank Probes
- The Trim Tank Transfer System Pipes
- The Trim Tank Vent System Pipes

Normally the Vent System Valves, the Vent System Pipes, Drain System Valves and the probes of the Trim Tank would belong to the Fuel Storage. However, as they are only necessary if a Trim Tank is foreseen they were related to the CG Control function.

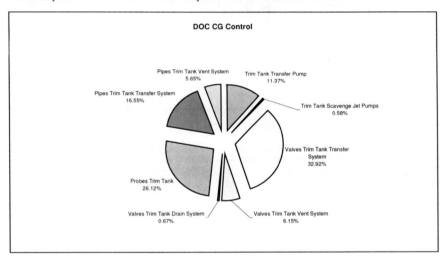

Figure 98: DOC Fractions of CG Control

6.5.6 DOC by Fuel System Functions

With this knowledge the breakdown for all functions was calculated as follows:

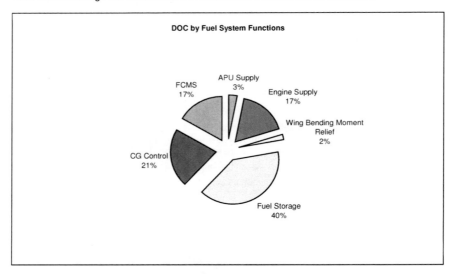

Figure 99: DOC Fractions of Fuel System Functions

As shown in Figure 100 the Fuel Storage causes by far the greatest fraction of the Fuel System DOC. The CG Control follows with 21% and the Engine Fuel Supply and the FCMS are equal with 17%. The APU Fuel Supply and the Wing Bending Moment Relief only contribute a very small fraction.

7 Conclusions & Recommendations

The algorithms of DOC_{SYS} have been incorporated into a DecisionPro™ analysis tree. This approach provides good visibility and easy modifications of the input parameters. To only evaluate the DOC of a whole system the tree structure is simple and can be completed in relatively short time.

However, to evaluate system DOC, it is very important to understand the influence of the individual system components on DOC. Thus it is necessary to calculate the DOC of the individual components. This approach can be done with DecisionPro™ but needs a more complex tree structure, as each component has to be evaluated with the DOC_{SYS} algorithms.

The DOC calculation for a Long Range Aircraft Fuel System Architecture resulted in one important conclusion. As shown in Figure 80, the maintenance costs are the main component of the Fuel System DOC. Thus any attempt to reduce Fuel System DOC should concentrate on its maintainability and supportability.

An investigation of the Fuel System component its DOC exposed the FCMS DMC as the main driver. At this point it is only known, that the FCMC are less reliable in comparison with other Fuel System components. Keeping in mind that the FCMS only consists of two computers a further investigation of these components is clearly needed.

The Inner Tank Probes are the second biggest DOC driver, with the maintenance costs cause the greatest contribution. The reason for the great DMC value is probably the great number of probes inside of the Inner Tanks and the fact that the related tank has to be defuelled if a probe is to be removed. Therefore a reduced number of probes would also reduce the DOC, although the impact on fuel quantity indication sensitivity would need to be understood.

A similar situation is given for the Trim Tank Transfer System Valves and Engine Feed System Valves, as these systems comprise a lot of valves and thus the maintenance costs of these systems are greater.

Regarding the DOC of the CG Control it would be interesting to compare this value with the Fuel Burn due to provide down force with the HTP.

The incorporation of DOC modelling from component level up to system level offers a very powerful causal link between the performance of the system and the performance of the components. The latter could be real, predicted or target data, thus offering powerful capability during the preliminary systems layout definition stages, where many competing options may need to be evaluated. Although modelling the system bottom up gives rise to more nodes and branches in the tree, the software enables very easy navigation, editing and execution of different data sets. Further work is needed to understand how changes in a baseline Fuel System due to introduction of new technologies can be reflected in the type of data needed in order to quantify the Δ in DOC due to that technology.

This study has required the linking and communication of data from many specialist sources in Airbus. The tree structures of DecisionPro™ can be written in html format on web pages to make them available for trans-national communities. Such an approach could provide a robust basis for down selection decisions to be made and by providing such clean traceability could prevent re-work in later project phases.

8 Summary

This thesis investigated the Fuel System DOC of a Long Range Aircraft with two engines.

The definition of operating costs and their contribution on LCC in general have been described. The DOC_{SYS} has been used to estimate DOC for aircraft systems and is also described. To calculate the Fuel System DOC with DOC_{SYS}, several system specific parameters such as price, weight, maintenance costs and power consumption were collected as input to DOC_{SYS}.

The DOC_{SYS} method has been used with the risk analysis tool DecisionPro™. A short example is shown to explain the basic functionality of DecisionPro™, and the use of Monte Carlo Simulation to evaluate uncertainties.

The Fuel System Architecture of a twin engined Long Range Aircraft has been described, and an overview given of which components comprising such an architecture are required to achieve the main functions of such a system. These functions include:

- Fuel Storage
- Engine and APU Fuel Supply
- CG Control
- Wing Bending Moment Relief

Also the parameters of the individual components have been collected and presented in non-dimensional form.

The algorithms of DOC_{SYS} were programmed into DecisionPro™. Several interactive tree structures were created in DecisionPro™ to calculate Fuel System DOC of the whole system, and the individual components.

The results of the calculations indicated that the maintenance costs are the main driver of the Fuel System DOC. The greatest component level contribution to DOC comes from the FCMS.

Monte Carlo Simulation has been used to evaluate OI costs due to the Fuel System. Stochastic inputs were made for the uncertain values of cancellation probability and cancellation cost. Monte Carlo Simulation provided statistical charts that indicate how these uncertainties the Fuel System DOC.

References

Airbus/University of Bristol 2002 AIRBUS; UNIVERSITY OF BRISTOL, *Design Project*. Airbus Industries, 2002

Bradshaw 2004 BRADSHAW, Philip: *Application of Monte Carlo Analysis to Quantify the Impact of Stochastic Data on System Cost of Operation*. Integrated Graduate Development Scheme, February 2004

Herinckx/Poubeau 2000 HERINCKX, Emmanuell; POUBEAU, Jean-Pierre: *Methodology for Analysis of Operational Interruption Costs*. FAST, September 2002

Hume 2004 HUME, Chris: *Brief note on future kerosene prices*. Airbus Memo, February 2004

Rogers 2002 ROGERS, Brian: *Easyjet – The Web's Favourite Airline*. International Institute for Management Development, 2002

Roskam 1990 ROSKAM, Jan: *Airplane Design Part VIII*. The University of Kansas, 1990

Scholz 1998 SCHOLZ, Dieter: *DOCSYS – A Method to Evaluate Aircraft Systems*. In: SCHMITT, D. (Hrsg.): *Bewertung von Flugzeugen (Workshop: DGLR Fachausschuß S2 - Luftfahrtsysteme, München, 26./27. Oktober 1998)*. Bonn : Deutsche Gesellschaft für Luft- und Raumfahrt, 1998

Scholz 2002 SCHOLZ, Dieter: *Aircraft Systems – Reliability, Mass, Power and Costs*. In: *EWADE '02 (European Workshop on Aircraft Design Education, Linköping, Sweden, 2. - 4. Juni 2002)*. Linköping University, Department of Mechanical Engineering, 2002